THE
LANGUAGE
OF BUTTERFLIES

How Thieves, Hoarders, Scientists, and Other Obsessives
Unlocked the Secrets of the World's Favorite Insect

WENDY WILLIAMS

SIMON & SCHUSTER

NEW YORK   LONDON   TORONTO   SYDNEY   NEW DELHI

Simon & Schuster
1230 Avenue of the Americas
New York, NY 10020

First Simon & Schuster hardcover edition May 2020

SIMON & SCHUSTER and colophon are registered trademarks of Simon & Schuster, Inc.

For information about special discounts for bulk purchases, please contact Simon & Schuster Special Sales at 1-866-506-1949 or business@simonandschuster.com.

The Simon & Schuster Speakers Bureau can bring authors to your live event. For more information or to book an event, contact the Simon & Schuster Speakers Bureau at 1-866-248-3049 or visit our website at www.simonspeakers.com.

Interior design by Ruth Lee-Mui

Manufactured in the United States of America

1   3   5   7   9   10   8   6   4   2

Library of Congress Cataloging-in-Publication Data
Names: Williams, Wendy, 1950– author.
Title: The language of butterflies : how thieves, hoarders, scientists, and other obsessives unlocked the secrets of the planet's favorite insect / Wendy Williams.
Description: New York : Simon & Schuster, 2020. | Includes bibliographical references and index. | Summary: "In this fascinating book from the New York Times bestselling author of The Horse, Wendy Williams explores the lives of one of the world's most resilient creatures—the butterfly—shedding light on the role that they play in our ecosystem and in our human lives"—Provided by publisher.
Identifiers: LCCN 2019025246 (print) | LCCN 2019025247 (ebook) | ISBN 9781501178061 (hardcover) | ISBN 9781501178078 (paperback) | ISBN 9781501178085 (ebook)
Subjects: LCSH: Butterflies]—Popular works.
Classification: LCC QL544 .W55 2020 (print) | LCC QL544 (ebook) | DDC 595.78/9—dc23
LC record available at https://lccn.loc.gov/2019025246
LC ebook record available at https://lccn.loc.gov/2019025247

ISBN 978-1-5011-7806-1
ISBN 978-1-5011-7808-5 (ebook)

For

LINCOLN BROWER
(1931–2018)

and to the memory of murdered activist

HOMERO GÓMEZ GONZÁLEZ
(1970–2020)

Nature has a perverse preference for the six-legged.
Michael S. Engel

Contents

Introduction   xi

Part I: Past   1

1. The Gateway Drug   3
2. Down the Rabbit Hole   15
3. The Number One Butterfly   27
4. Flash and Dazzle   45
5. How Butterflies Saved Charles Darwin's Bacon   67

Part II: Present   77

6. Amelia's Butterfly   79
7. A Parasol of Monarchs   87
8. The Honeymoon Hotel   97
9. Scablands   111
10. On the Raindance Ranch   121
11. A Sense of Mystical Wonder   135

Part III: Future   143

12. The Social Butterfly   145
13. Paroxysms of Ecstasy   171
14. The Butterfly Highway   181

Epilogue: In the Mountains of Mexico   199
Acknowledgments   203
Notes   205
Photo Credits   223

Introduction

Color is a power which directly influences the soul.

Wassily Kandinsky

Long ago, when I was twenty, penniless, and hanging in London, looking for something free to do, I drifted into the city's Tate Gallery—filled with some of the world's best-known art—and walked straight into a staggering J. M. W. Turner masterpiece.

I was gobsmacked.

Knocked for a loop.

Brilliant and shimmering, shrieking with yellows and oranges and reds swirling around smoky-vague outlines of battling ships at sea, that painting owned me.

If you've seen Turner's creations, you know why. His works tap into a secret crevasse in the human psyche, a down-the-rabbit-hole neural pathway from which, for some of us, there is no escape. It's a biological thing. An evolutionary mandate. Only recently discovered by science but long understood intuitively by artists, this hidden desire elicits a unique kind of hypnotic trance—a craving for color.

Standing before Turner's work, I was mesmerized.

I tried to wind my way through the mysteries of the thing. This was pure experience. I knew nothing about Art. I was an innocent. I had no idea who Turner was, no clue that he was considered a genius who paved the way for Impressionism. I had not been prepped to venerate his work. This was a once-in-a-lifetime thing.

A first kiss.

I was never again so deliciously, so exquisitely, so naively shocked.

Until . . .

I was sucker-punched once again. This time it was in Larry Gall's Yale University offices. Interested in the crazy, titillating, and sometimes even deadly world of butterfly obsession, I had come to meet Gall, the university's trim, bespectacled computer expert and keeper of more than a century's worth of butterfly, moth, and caterpillar collections. Brought to Yale from locations worldwide were thousands upon thousands of boxes of carefully pinned and lovingly recorded specimens of Lepidoptera—butterflies and moths.

Like the Turner, these boxes were monumental works of art. But unlike the massive Turner seascapes I'd loved, these boxes had been squirreled away for decades in hundreds upon hundreds of protective climate-controlled drawers. Assembled by compulsive butterfly addicts who worked in isolation in rooms and jungles and labs worldwide, some boxes dated to the eighteenth century.

The artists who made them obviously combined a deep passion for color with a meticulousness for detail. These kaleidoscopic assemblages represented hundreds of lifetimes of devoted labor by men and women hunched over their desks, working with a steadiness of hand and of intention that I could only dream of.

More than four decades after my life-changing love-at-first-sight view of a Turner, I was stunned all over again. I wanted to see more.

And more.

There was a lot to see. Yale has literally hundreds of thousands of butterfly and moth specimens. The boxes are cosseted in drawers that run from floor to ceiling, in line upon line of cabinets, in the expectation that someday, somewhere in the universe, whether in our own Milky Way or beyond, some researcher, perhaps yet unborn, will need them for a study.

Neatly pinned in meticulous rows, entire trays were devoted to only one species. The best of these boxes also note when and where the samples were collected.

Gall patiently pulled tray after tray after tray of butterflies. Just as with Turner's painting, I struggled to make sense of what I saw. Who knew that trays of dead insects could be so delicious, so sensuous, so entirely luscious?

Eventually Gall, himself an addict, wearied of me and my incessant "Why this?" and "Why that?" I was politely, gently, but definitively dismissed.

And so I learned that the butterfly effect (to repurpose a term) is real, that this craving for color so deeply hardwired in our brains could turn into an addiction. What had been an arms-length inquiry into the unusual desires of certain lepidopterists had exploded into a compulsion of my own: What exactly *were* these odd flying creatures, some so small as to be almost invisible but others with foot-wide wingspans?

Like most people, I was no stranger to butterflies. Butterflies had been my companions for most of my life, as I rode horses through high Rocky Mountain valleys or over rich wildflower-filled Vermont fields. They were common in the Pennsylvania meadows where I grew up, as they were when I lived in Senegal or traveled in Zimbabwe or Kenya or South Africa. Everywhere I walked among weeds and wildflowers, butterflies fluttered. As I hiked Appalachia's mountain trails, or strolled the beaches of Cape Cod, butterflies were there.

Of course I had seen them. Of course I liked them. Who doesn't? But I took them for granted. I hadn't really *looked* at them. Not closely, that is. Where did they come from? Why were they here? What the heck do they get up to while they're on our planet? And what is it about them that compels the human psyche so insistently that men and women have risked their fortunes and their lives and, on occasion, died in order to capture them?

My curiosity was about to take me around the world—sometimes literally, sometimes by reading or talking on the phone with a multitude of scientists who knew exactly what I meant when I told them about my lepidopteral epiphany. As the veil lifted from my eyes, an entire universe opened to me.

I learned that the language of butterflies is the language of color.

They speak to each other using that flash and dazzle. I sometimes imagine them as the world's first artists. Happily for us, humanity finds joy in that same language of color. We have an ancient partnership with these six-legged life-forms that has helped us survive throughout our 200,000-year planetary presence.

Butterflies continue to partner with us even today. I learned that the seventeenth-century study of butterflies revolutionized our understanding of nature and thus provided the foundation for the field of scientific research we now call ecology. I also learned that this foundation was laid down by the research of a highly methodical, meticulous thirteen-year-old girl.

I learned that unlocking the secrets of butterflies helped us understand how evolution works, that their partnership with other living things forms the basis of life on our planet, and that butterflies today are helping us in many practical ways, improving our own lives by providing surprising new models for medical technology. For example, butterfly scales are helping materials researchers biodesign devices to help asthma sufferers.

All these surprises whetted my curiosity. When I started this project, I thought that writing about butterflies would be a simple matter. I was wrong. Butterflies are wonderfully complex beings that have evolved for well over 100 million years. Excitingly, while we have recently made great strides in unlocking their secrets, some of their unique attributes have yet to be understood.

Sadly, I also learned that, for a multitude of reasons, butterfly and moth population numbers are dropping, sometimes precipitously. There are many reasons for this decline and many actions that can be taken to prevent further losses. I learned that the disappearance of butterflies would be a planetary disaster, and not just for esthetic reasons. Their essential services keep the entire system intact.

Luckily, science has already achieved a great deal when it comes to butterfly conservation—so there is hope for the future. Hundreds of researchers worldwide and thousands of dedicated butterfly groups are making a difference.

In this book, we'll find out how.

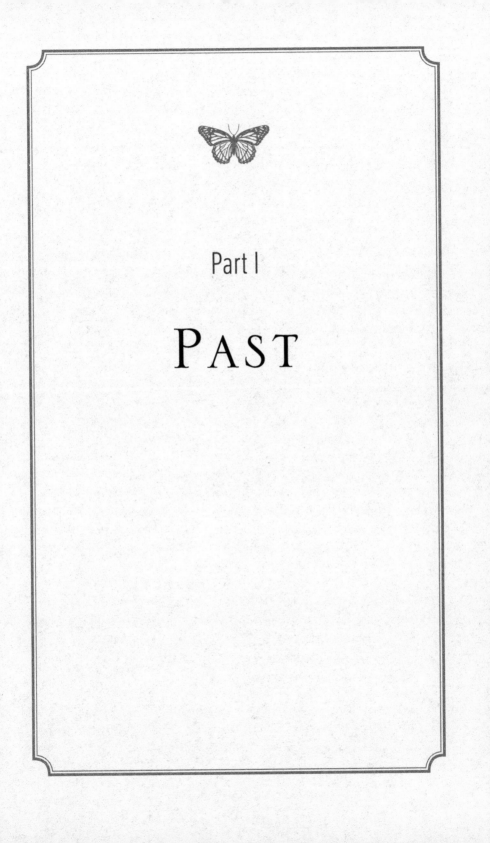

Part I

PAST

One

THE GATEWAY DRUG

A lepidopterist will be as familiar with the speckles and dappling of a butterfly wing as he would be with the faces of his own family. One lepidopterist I knew was actually rather more aware of the former than he was of the latter.

Richard Fortey, Dry Storeroom No. 1

Herman Strecker was, by all accounts, a very odd man. He had a long face and a long neck and an even longer, out-of-control beard. He looked like Moses. He had deep-sunken grief-filled eyes. He lived the unkempt life of a zealot, going so far as to crawl in between his bedsheets with his pants and boots on.

By day, he was a poor stone carver who specialized in carving angels on children's gravestones. But by night, Strecker descended into a deeper, darker lust—a greedy compulsion that eventually dominated his entire existence. Some people want to possess money. Others want to possess clothes or cars or stamps or houses or politicians.

Strecker wanted butterflies. Lepidoptera. (That's Latin for butterflies and moths, *lepidos* being the Greek word for "scale"; more about this later.) He yearned to own at least one specimen of every butterfly species on Earth. He came close. By the time he died in 1901, having lived a life of intense emotional desperation, he had amassed 50,000 specimens. I can't imagine having that many of any one thing in my home. There must have been precious little room for anything else.

That's a small number compared to British banking scion Lord Walter Rothschild's 2.25 million. Lord Walter, active at the same time, was one of the planet's wealthiest men; he had special facilities built to house the collection and employees to look after them. Strecker was most definitely not among that 1 percent. Nevertheless, Strecker's collection was then North America's largest. Given his extreme poverty, I would imagine that pinned, dead butterflies must have been stashed throughout his not-particularly-large abode.

Strecker was a product of his Victorian world. Indeed, he died in the same year that Queen Victoria herself died. His tragic life was filled with dead babies and deprivation and women who died young and hunger and an acerbity so extreme that his tale sounds straight out of Edgar Allan Poe. In fact, the gravestone carver actually did sculpt a raven for the entryway of one Philadelphia client's mansion, which seems fitting, given his character. Like the lover in Poe's "The Raven," slowly descending into madness, Strecker was a feverishly despondent man. The older he got, the more extreme he became.

He was "omnivorous," he once wrote. Never satisfied, like Midas after gold. "My soul pines," he told a friend, when seeking an exotic butterfly that proved difficult to acquire. When another person sent him a long-desired birdwing butterfly, he wrote: "There is no use trying to express my feelings at beholding the splendid ornithoptera. Only to think the dream of my childhood fulfilled for since I was five years old I coveted and fretted for the Green Ornithoptera." And in yet another letter, he asked: "Why did God implant in us unquenchable desires, and then deny the means of gratifying them?"

As a child, Strecker had once been permitted to look at some expensive hand-painted books on butterflies in a Philadelphia natural history museum. In the early 1800s, northern cultures were monochromatic. Cities and towns were covered in soot and grime from woodsmoke and coal smoke. Even people, save for the ultrarich, wore blacks and grays. The world of print, too, was colorless.

These hand-painted books, in contrast, were remarkable in their lavish grandeur, with depictions of exotic butterflies that lived in faraway tropical countries. They were the early-Victorian equivalent of today's grand epic films.

I imagine Strecker, the child, was as overwhelmed by those books as I was by the Turner. Into his drab world of soot, poverty, and hopelessness, the goddess of color had made her debut. He began netting butterflies near his home, pinning them to boards in order to preserve them. The infatuation infuriated his father. A slew of paternal beatings followed, but Strecker would not—or perhaps *could* not—give up his obsession with beauty and sunlight.

Strecker was not alone. During the Victorian era, collecting and naming God's creatures was an approved endeavor shared by all classes of people. Even women were allowed to play. Throughout Europe and North America, insect collecting was considered not only a healthy activity, but a way to honor God and His earthly works—and was therefore acceptable even in dour cultures where play was frowned upon.

Indeed, humanity had a "*duty* to inventory," writes paleontologist Richard Fortey in *Dry Storeroom No. 1*, a personal memoir about treasures that lay sometimes helter-skelter in the back rooms of London's Natural History Museum even to this day.

That "duty" was based in Biblical texts. In Genesis, the Victorians read that God formed every living thing on Earth and then commanded Adam to name them. Before they could be named, of course, they had to be collected.

"Collecting was a Victorian passion," writes Jim Endersby in *Imperial Nature*. "From shells, seaweeds, flowers, and insects to coins, autographs, books, and bus tickets, Victorians of every class collected, classified, and arranged their treasures before exchanging unwanted finds with other enthusiasts." (*Bus tickets?*)

This gave rise to the enjoyment of being outdoors just for the joy of being outdoors, of having what American Victorian poet Walt Whitman called "a butterfly good-time." But for some, the collecting addiction went far beyond mere cultural expression, so much so that it might well have had a genetic basis.

In the last decades of the nineteenth century, the most proficient butterfly collectors—and there were many—knew each other. They corresponded regularly. Strecker, widely acknowledged as North America's foremost expert, was part of that club. Eventually, though, other collectors began to suspect that when Strecker visited their collections, he departed with a purloined specimen or two. He was increasingly snubbed.

He became vituperative. He lashed out at colleagues, who returned fire. One called Strecker an "entomological spider." In 1874, a collector and one-time friend, in what came to be called "the Central Park Affair," accused him of stealing specimens from what is now the American Museum of Natural History. The accuser enjoyed high status in the butterfly world. He was widely believed.

The claim went like this: Strecker wore an Abe Lincoln–style stovepipe hat. Inside, rumors asserted, was a hidden corkboard on which he pinned his purloined samples. This was never proven. Still, many museums would not allow him to visit their collections. No evidence of his criminality has turned up in the century since his death. It's possible that he was accused because of his unusual nature. The depth of his passion may have isolated him from his colleagues.

Strecker died a bitter man. His collection is now in Chicago's Field Museum, along with 60,000 letters and books, testimony to a lifetime's dedication or addiction, depending on your point of view.

Strecker's biographer and the author of *Butterfly People*, William Leach, calls Strecker an "antinomian" (rule-breaker) of the butterfly world. Leach believes that Strecker was not guilty of the theft, but that his belligerent nature kept him from mingling well with other collectors, many of whom were from the wealthier classes. We talked on the phone and discussed whether Strecker's yearning to collect butterflies might have involved a genetic predisposition.

"I have the same gene," Leach told me. "I perfectly understand the man. It just overtakes them. It's an unexpected kind of thing. It starts with the initial encounter of the child with this flying color. It produces something in the child: I want that. I *want* that."

But that, warned Leach, is only the beginning.

The more you learn about butterflies, he said, and then moths—Lepidoptera—the more infatuated you are.

"Butterflies," I was warned by several researchers, "are just the gateway drug."

Down the rabbit hole.

So what is it about butterflies that so easily and so universally catches the fancy of Earth's *Homo sapiens*? Is it merely that they are pretty little things? Or is it perhaps in part that they are a symbol of our planet's always evolving story, a symbol of our partnership with all other living things, a symbol of the circle of life?

There may be as many as 1 trillion total species living on Planet Earth. Most remain undiscovered. Somewhere around 1.2 million species have been named and formally described. Given that Victorians only seriously began the task of naming all living things a little less than 200 years ago, that's pretty good progress. But it will be many, many lifetimes before we truly have a handle on *all* species on just our own planet. And who knows what's in the universe beyond our own tiny world? Molecular biologist Christopher Kemp sums it up this way: "How little we know about the natural world that thrums and vibrates all around us."

By far the majority of Earth's species are single-celled living things, both with and without a nucleus (the central structure in the cell housing DNA). But most people think in terms of plants and animals. Most animals are multicellular and mobile; most plants are multicellular but not mobile. (Although, of course, there are exceptions to the rule.)

Of plants, fewer than 400,000 species are known. Compare that to the number of named insects: currently roughly 900,000. Compare that to the number of known mammal species: about 5,400.

Ergo: Insects rule.

"Evolution begets diversity," write entomologists David Grimaldi and Michael Engel in *Evolution of the Insects*, the go-to text for insect scientists. Since insects have been around for hundreds of millions of years—certainly longer than any mammals—and since many insect species have survived the planet's relentless extinction events, it stands to reason that there would be a plethora of them.

An insect is a kind of arthropod, a being with an external skeleton. Its ancestry dates all the way back to the glamorous world of the Cambrian, when evolutionary experimentation ran amok and the sudden richness allowed life to explode in the seas. Beginning about 540 million years ago, arthropods ruled. They were the best idea around.

As arthropods, butterflies can trace their roots all the way back to this time, long before animals with skeletons on the inside were common. "By most measures of evolutionary success, insects are unmatched: the longevity of their lineage, their species numbers, the diversity of their adaptations, their biomass, and their ecological impact," write Grimaldi and Engel.

Insects have been around for 400 million years. By contrast, the most primitive mammals seem not to have appeared until about 140 to 120 million years ago—around the time of the first flowering plants. We have no solid evidence that modern mammals, such as primates and horses, existed until about 56 million years ago. It is indeed as the great population scientist E. O. Wilson says: Little things run the Earth.

"Without a doubt," write Grimaldi and Engel, "the diversity of any other group of organisms has never been more than a fraction of that of insects." Other than single-celled organisms, of course.

So how do butterflies fit in? They belong to the second-largest order of insects currently extant: Lepidoptera—insects with scales on their wings, including about 180,000 known species. (There are probably a whole lot more yet to be discovered and named.) Of those, only about 14,500 are butterflies. This figure reaches about 20,000 if you include a group of insects commonly called "skippers," which some scientists classify as butterflies and some do not.

The other 160,000 or so flying insects with scales on their wings are called "moths." What exactly, I wondered, was the difference between moths and butterflies? How is it that they are the same—but different?

In a lab at Yale, I talked about this with some volunteers who were helping organize the university's extensive collections. The word "moth" elicited distaste. We made the classic "disgust" face while we talked about them: crinkled-up noses, slightly flared nostrils, and pulled-back, almost snarling lips. When we talked about "butterflies," eyes lit up and smiles appeared. There is even an official name for our dislike of moths—*mottephobia*—while as far as I know, there is no official word for "fear of butterflies." Many people who fear moths find butterflies delightful.

In our discussion, the two groups of Lepidoptera evoked distinct emotional responses. "Moths" were annoying and sometimes costly invaders that infested your baking flour and ate up your woolens and bothered you by flying around your electric lights at night. "Butterflies," on the other hand, were whimsical, delicate, pure, virtuous, clean, in need of protection, ornaments that accentuate the beauty of your garden's flowers.

These are prejudices. Not all cultures find moths repulsive. Some people enjoy them. Others are sustained by them. Aboriginal Australians traditionally hunted for large populations of semidormant bogong moths, which they then roasted and either ate immediately or ground up into a portable, edible protein paste that they could conveniently carry around with them, like pemmican.

Other cultures find moths useful in other ways. In Taiwan there are populations of the Atlas moth, or "snake's head moth"—so called because when threatened the flying insect drops on the ground and writhes slowly, so that its wing tips look like the head of a writhing cobra. Female Atlas moths have a wingspan of up to twelve inches. When the Atlas moth emerges full-grown from the cocoon (moths emerge from "cocoons," butterflies from "chrysalises"), local people use the silk-based, now-empty envelope as a purse.

I had never seriously thought about the differences between moths and butterflies. It had just seemed obvious to me. I decided to find out more.

• • •

In the butterfly collection at Harvard's Museum of Comparative Zoology, curatorial assistant Rachel Hawkins walked me over to a box where several specimens were pinned. At only a few hundred thousand Lepidoptera, the collection is small compared to Rothschild's, but is nevertheless eminent in that it contains Lepidoptera collected by a man later eaten by cannibals and a huge birdwing butterfly hunted by shotgun. This specimen was likely collected by one of the museum's earlier directors, the antievolutionist Thomas Barbour, who as recently as World War II firmly believed that evolution and genetics were not connected.

"Tell me which are moths and which are butterflies," Hawkins said.

In the box were eight specimens, arranged in two columns. In the top left-hand column was a large insect with iridescent wings, brightly colored in greens and yellows, and slender-bodied. It was dazzling. Next to it, in the top right-hand position, was a thick-bodied, clumsy-looking insect with a bloated abdomen that reminded me of a rather large and wicked-looking bee. The wings were mostly dark, with thin streaks of yellow. I guessed that the top left-hand insect was a butterfly, because of its slenderness and colorful wings. I guessed that the top right-hand insect was a moth, primarily on the basis of the thick body.

And so I went on, through the whole of the specimen box, using the rules of thumb I had been taught: Moth antennae are thick and hairy, while butterfly antennae are slim and slightly knobbed at their tips. Moth bodies are chunky while butterfly bodies are sleek. Moths fly at night and butterflies fly during the day. Moths are dull-colored; butterflies are beautiful.

Or so claims the common knowledge.

I was wrong every time.

Hawkins told me: "People think that moths are very drab things, little brown jobs that come to your light at night and they all look the same. That's definitely not the case. There are plenty of moths that have bright coloration, and there are butterflies that are just little brown cryptic things."

There are also, she continued, plenty of moths that fly by day and some butterflies that fly at dusk.

"People often look at body shape and characteristics," she continued, "thinking that moths are stout and chubby and furry and butterflies are not. That's definitely not true. Some of the butterflies that are more powerful fliers will have thicker bodies. There are certainly moths that are slender and graceful, even some that imitate wasps' slender bodies." Moths usually appear "fluffy" while butterflies appear sleek, but swallowtail butterflies are also "fluffy," possibly because they can fly at high altitude and it's colder up there. They need protective insulation.

It wasn't the first time moths had confused me. One day shortly after I began working on this book, I looked out my living room window at my favorite shrubs that attract butterflies. I saw what seemed to be the smallest hummingbird I'd ever seen. In Cuba I'd once been fascinated by that island's bee hummingbird, *Mellisuga helenae*, the world's smallest bird and indeed about the size of a very (*very*) large bee, definitely *not* the kind of bee I'd want to meet in my flower garden.

My first and rather irrational thought was, "I wonder how that tiny bird has managed to fly all the way north to Cape Cod from Cuba?" I watched for quite a while. The hungry thing hovered over flower after flower, seeming to take a sip and then hover elsewhere, seeming to take another sip.

The more I watched, though, the more suspicious I became. This was not the behavior I'd expected to see. There was too much hovering going on. And not enough flitting. Hummingbirds are notorious for refusing to stay still, much to my frustration, since I love watching them. This little thing was just a bit too calm, staying on the same bush and moving almost methodically from flower to flower.

My eyes narrowed. I squinted, just to be sure. I had been fooled. This was no hummingbird.

This was a hummingbird *moth, Hemaris thysbe*. This guy was flying during the day, just like hummingbirds and butterflies. He was reddish-colored and stood out like a sore thumb on my purple bush. He did have a chunky body, but he was beautiful.

Some moths have evolved to look much like butterflies. But the Madagascan sunset moth *behaves* like a butterfly in many respects.

When first named in the late 1700s, the sunset moth was classified as a butterfly, in part because it flies during the day rather than at night and because it is extremely colorful.

There is one fairly tried-and-true method for differentiating between moths and butterflies: the *frenulum*. Moths have them on their wings; butterflies do not. (Of course, there are exceptions.) Essentially, the frenulum is a hooking system. Moths have, on both sides, a forewing and a hind wing. The front and hind wings move in unison because they are hooked together. The formal terminology for this system is *frenulo-retinacular wing coupling*, but it's easiest if you simply think of it as a kind of hook-and-eye system.

Butterflies do not have this system. Instead, they usually have powerful and large forewings, which, when flying, cover so much of the hind wings that the forewings simply push down the hind wings with what might most easily be described as brute force. (But again, there are exceptions to this rule. When you've been evolving for tens upon tens of millions of years, there are always going to be exceptions.)

On the other hand, moths and butterflies do share some important characteristics, including a remarkable proboscis. Pronounced "pro-BAH-sis," this weird word has a simple meaning: "long snout." Elephants have outstanding proboscises. So does Taff, my border collie, whose long proboscis is always out there, snuffling beneath leaves and surveying the ground for the presence of sheep or bad guys or girlfriends as we walk. The aardvark, an African mammal, has a burdensome-looking proboscis that sniffs out ants and termites. The proboscis monkey has a fantastic snout, although no one knows why.

But moths and butterflies have distinctive "snouts," fantastical appendages that are *not* noses and are *not* used to take in oxygen and are *not* used to sniff things out. (To breathe, Lepidoptera have tiny holes called *spiracles* in their exoskeletons that take in oxygen. They detect aromas via their antennae.)

Proboscises take in nourishment without chewing, slurping, lapping, or licking. Sometimes people liken lepidopteran proboscises to "tongues," but that's not truly the case, since a tongue sits in a mouth and neither butterflies nor moths have "mouths" in an everyday sense. Sometimes proboscises are described as "mouthparts," but that's just convention.

Lepidoptera proboscises are bizarre, outlandish, and even at times slightly grotesque, extensions of the insect head. They are unlike any other organ with which most of us are familiar. Lepidoptera proboscises are sometimes three, or four, or even five times the length of the insect's body.

Only flying-stage Lepidoptera have these prodigious appendages. Caterpillars, eating machines that they are, have mandibles—jaw-like hardened parts of the exoskeleton operated by muscles—that are always at work, mashing up food and storing nutrients and toxins for use in their coming incarnations as flying insects. ("Chewing" is not truly accurate in the conventional sense, since they don't have teeth.)

Inside the chrysalis, as the caterpillar changes into a butterfly, the mandibles disappear. The muscles that operated the mandibles dissolve in a bath of nature's corrosive chemicals, a.k.a. "enzymes." (Of course, there are a few moths that still have mandibles when they emerge. Exceptions, exceptions, always exceptions.)

At the same time, other clusters of cells become active, producing, among other organs, the lepidopteran proboscis. The proboscis develops in the chrysalis as two separate elongated halves of a tube. When the butterfly emerges, the two halves, each with its cross section in the shape of a C, come together to form an elongated O. This elongated O, this tube, may be only a few millimeters in length, or it may be considerably longer.

Lacking "mouths" to accomplish refueling, most Lepidoptera use their proboscises. This nourishment-acquisition instrument curls and uncurls repeatedly over the animal's life span, an infinite number of times. Think of the paper horns children relentlessly blow and re-blow at parties.

The proboscis does just what its name implies: It probes. It explores. It hunts for food. If you sit quietly and watch closely, you can see a butterfly at a flower, probing the interior of the flower, looking for nectar.

Normally, when the butterfly is simply flying, the proboscis is curled coil inside coil, like the tubing of a French horn. But when it's time to uncurl the proboscis in order to do this exploring, two sets of muscles—one on each side of the curled tube—flex, stretching the tube out its full length, somewhat the way an elephant stretches out her trunk.

If you've watched a butterfly on a flower for any length of time, you've seen that with this uncurled tube they seem to be sipping from a pool of nectar that we imagine lies in wait inside the flower. (We'd be wrong, but we'll get to that.)

The proboscis is where the rubber meets the road. Where insect and flower join together in joyful partnership. It's a marriage not just of convenience, but of sustenance. Flowers, with their alluring scents and sweet nectary, tempt the insects to come hither. The insects, while obtaining nectar (or "nectaring," in lepidopterist parlance), inadvertently also obtain pollen, which they obligingly but unintentionally carry to the next flower, so that the flower is fertilized by a new set of genes. The insects do not *plan* to facilitate flower sex, but that's exactly what they do.

Because of the proboscis, the insect both gets something and gives something. Ditto for the flower. It's a mutual exchange. It's also essential, if life is to survive on our planet. Knowledge of this fact is something we take for granted these days, but for most of human history we did not understand this simple natural truth.

Until the early 1800s, Western thinkers explained flowers as God's gift of beauty to humanity. Their purpose on the planet was to thrill us and thus impress upon us the presence of God in our lives. We can still think of them that way, certainly, but about 200 years ago horticulturists grasped a different level of truth: that flowers used sex *(sex!)* to reproduce. Flowers had male parts and female parts, and pollinators helped make the connection. Sex! The idea was so horrifying that it could not be discussed in the presence of women or children. But eventually, truth will out. We came to accept the awful fact of life: butterflies (and other insects) provide an important pathway for this sexual exchange of genes.

Indeed, this relationship between flower and proboscis eventually inspired a central insight into how evolution works.

Two

Down the Rabbit Hole

Something as simple as a butterfly contains complicated mysteries
that you and I will never understand, and that—that's beautiful.

Destin Sandlin, Smarter Every Day

At the end of January 1862, when Herman Strecker was in his mid-twenties, Charles Darwin, nearing his fifty-fourth birthday, was
writing a letter to his best buddy, botanist Joseph Dalton Hooker. He
was expounding on his dismay over the American institution of slavery,
which he passionately abhorred. Then he complained that the British institution of primogeniture —the law requiring that estates be inherited
by the oldest son—was also problematic, in that it interfered with the
law of natural selection: "—suppose the first-born Bull was necessarily
made by each farmer the begetter of his stock!"

At least fifteen members of his household (the quintessential family
man, he had lots of kids and many servants) and Darwin himself were
recovering from a nasty flu. Nevertheless, he was hard at work on a se-
quel to his 1859 best seller, *On the Origin of Species.* He had high hopes
for his new soon-to-be-released tome: *Fertilization of Orchids: On the
Various Contrivances by Which British and Foreign Orchids Are Fertilized
by Insects and on the Good Effects of Intercrossing.* (Oh well. In those days
it was thought that titles should tell readers exactly—*exactly*—what they
were paying for. No cheapening clickbait allowed.)

Writing to Hooker was Darwin's way of escaping the tribulations of an infected household and of unwinding after the hard work of creating his book. But his friendly ramblings, full of gossip and quips, were interrupted by the delivery of a package. We learn of this momentous incident also, jotted in haste at the bottom of the letter. Thank goodness Darwin loved to write letters.

The package contained a generous gift, something rare and precious—a spectacular six-petaled, star-shaped orchid, a marvelous flower native to Madagascar. Although he didn't know it when the parcel arrived at the door of his chaotic household, the gift would play a major role in his forthcoming *Orchids* blockbuster. It would be an important reason why the book, today nicknamed *Contrivances*, would continue to be read well into modern times.

Darwin's eyes opened wide not because of the flower itself, but because of the length of the appendage that hung down from its base.

It was huge, almost a foot long.

This enormity shocked Darwin.

He asked a question which, in various iterations, has kept scientists busy over the last 150 years.

"Good Heavens what insect can suck it" he wrote in a postscript to Hooker. He dashed this off in such excitement that he did not put a question mark at the end of the question.

The orchid was "astounding," he commented. Later he referred to the green "whip-like" foot-long spur, at the bottom of which, he believed, was stored the flower's nectar. If you've spent any time at all looking at orchids, you've seen such a spur. If you break open this spur, you'll see that it's hollow inside.

Darwin pondered this orchid. Why would a flower use so much energy to produce something that would make the flower's nectar so inaccessible? It didn't make sense. Wouldn't this deter pollinating insects, and thus limit the flower's reproductive options?

Finally Darwin realized: the flower wanted to lure not just any insect, but a particular, specific insect that wouldn't *misdeliver* the flower's pollen by carrying it off to the wrong flower species. By lengthening its spur, Darwin reasoned, this particular orchid was making itself alluring only to an insect species with an equally long proboscis.

It was a made-to-order situation, a hand-in-glove kind of thing. If the glove didn't fit, you wouldn't want to wear it.

And, he grasped, the insect, too, would benefit: it would be able to access the nectar without having to compete with other insect species. In other words, Darwin came up with a theory of pairing: with the theory, not just of evolution, but of *co*evolution, a theory that suggested a natural partnership among living things. In a win-win marriage, organisms sometimes evolve together in a reciprocal relationship.

Various living organisms, which we humans perceive as separate entities, sometimes fit together so well that they are almost one living thing. They *need* each other in order to survive.

Indeed, our entire planet can be thought of this way. This was not Darwin's own unique idea; others, beginning with Maria Sibylla Merian in the 1600s (more about this long-ignored genius housewife later), had begun to show nature as a net of living things. But Darwin verbally codified this idea, giving it an important solidity.

He proposed that eventually a Lepidoptera species would turn up that had an exceptionally long proboscis, a colossal organ that fit all the way down into the lengthy spur. He put his prediction in his forthcoming orchid book. Later on, he wrote that he was ridiculed for doing so. Few people could imagine a butterfly with such a pendulous proboscis. How would the insect fly having to carry such a thing?

For the remainder of his life, Darwin hoped some collector on Madagascar would find his predicted insect.

It was not to be.

At least, not in his lifetime.

It was the wealthy banker Walter Rothschild, he of the massive butterfly collection, and Rothschild's employee, entomologist Karl Jordan, who in 1903 affirmed Darwin's prediction, describing and naming the long-sought insect, a member of the family of hawkmoths, or sphinx moths. The specimen had been sent to them by two French field entomologists.

The body of the moth was not huge, but the appendage was, as predicted, almost a foot long. This seemed to be an affirmation, but there was another hurdle to jump: no one had actually *seen* the moth insert the proboscis into the orchid's spur.

It wouldn't be until our era, in the 1990s, that a field entomologist would film the behavior in the wilds of Madagascar.

So Darwin was right.

But only to a point. His story was nice, neat, and pretty. But it now turns out that, to get the full story of this marvelous partnership, his understanding of the proboscis needed some fine-tuning. The moth and the orchid were indeed a match, but the moth was not "sipping" nectar from the orchid. At least, not as he had envisioned the "sipping."

Not by a long shot.

Fast-forward to the end of the twentieth century, more than a hundred years after Darwin wrote that letter. Four-year-old Matthew Lehnert wandered into his parents' Michigan bedroom one day and saw a huge moth strolling across the pillows on his parents' bed. She was hard at work laying eggs.

Though he was barely out of toddlerhood, Lehnert's future was laid out before him at that instant. His fate was sealed: he would become an entomologist. Down the rabbit hole.

Just in case people didn't think he was serious about his career goal, on the Halloween of his sixth year of life he wore a white lab coat. On the back was written, in large letters: "Entomologist." Just to clarify.

When he got older he worked in entomological research labs, then studied the Homerus swallowtail, endemic to Jamaica and the Western Hemisphere's largest butterfly, now critically endangered. Next, Lehnert took a two-year position in the lab of a chief scientist who was studying the proboscis.

He wondered: What's to learn? It's a straw. The butterfly sips. Pretty simple, he thought, just as Darwin had thought more than a hundred years earlier. A pump in the butterfly's head sends the nectar, ultimately, to the insect's gut. Should take two months, he thought. What will I do for the rest of the time?

Ten years later he's still at it. In fact, he now has a lab of his own and minions who are as excited about proboscises as he is. The problem is, the organ is *not* what it seems: a simple drinking straw. Or rather, it is, but not exactly. . . . Many people, including highly accomplished entomologists, describe the proboscis as a tool with which the insects "drinks."

But the more accurate word would be "absorbs."

The proboscis turns out to be basically a highly sophisticated paper napkin.

For one thing, the proboscis is not airtight from tip to tip, as a straw is. "The proboscis is actually porous," he told me when I called him. "Try sticking a lot of holes in the straw and then sucking. It wouldn't work well. The proboscis turns out to be much more like a sponge."

I envisioned a kitchen sponge. If you put your hand on a sponge and squeeze, then put it in a sink full of water and open your hand, releasing the pressure, the sponge soaks in the water as it expands. An organ that pumps or sucks is not necessary: even if you just lay the sponge down on a film of water on your kitchen counter, the sponge will still soak up the water.

That's the idea, Lehnert confirmed. To ingest something, the insect lays down the proboscis *atop* the substance. Think of laying a paper towel over a spilled liquid. Without any effort on your part, the towel will blot up the liquid. That's how the proboscis works. The infinitesimally small holes in the proboscis absorb the substance and *voilà!* It's inside the proboscis, in the transport tube. No inhalation required.

It does this by using a technique we all learned about in grade-school science: capillary action. I remember learning about capillary action in third grade. It seemed like magic to me. By that age I'd learned that gravity pulls things down. Not up. And I'd learned the basic cause-and-effect patterns of life on Earth. Things don't just float up, defying gravity, by themselves. Even a kite needs wind and someone to hang on to the string.

And yet: my teacher stuck a thin glass tube into a beaker full of water. And, lo and behold, the water climbed *up* the insides of the tube. Along with the rest of my third-grade colleagues, I gasped: *That can't happen!* The teacher explained air pressure, and that the greater the air pressure, the higher the water would move. This made sense, and I could go back to my basic theory that, with gravity in action after all, life made sense. (Little did I know, but that's a different book. . . .)

Capillary action is a major player on our planet. Capillary action is why you can dry your dishes with a tea towel. Capillary action moves water from plant roots to leaves. Without it, we would have no sequoia trees, for example.

The same force moves liquids inside a flower or water from a pool on the ground to inside the butterfly's proboscis. Physically "sipping" is not necessary. In fact, no physical effort is required at all. Because the holes in the proboscis are so small, the liquid easily moves into those holes. Just as you saw in science class that water slowly climbs up the walls of the pipette, rising high above the level of the water in the jar, the small holes in the butterfly proboscis move liquid from a pool on a surface into the proboscis interior.

Ingenious.

However—and this is very cool—it's not just liquids that butterflies ingest this way. The butterfly can take up *dried* substances. Anyone who has ever gone for a walk in the summertime has seen a butterfly on a seemingly dry area, like a trail or a sidewalk or a rock, hard at work seeming to consume something. But how? There doesn't seem to be anything wet there. What's the deal?

There most likely *is* something there, invisible to us but obvious to the insect. Something that gives off a scent. It could be something like a thin film of salts left from urine deposited by a fox or coyote or dog. We may be oblivious to this precious substance. Through highly sensitive antennae, the insect detects it easily. But it's dry. How to take advantage of it?

Researchers have discovered that the insect lays the proboscis atop the film, then sends saliva *down* the proboscis tube and out through the tiny pores. The saliva suspends these salts in the liquid, which is then absorbed back into the proboscis and brought back up the tube. It's a two-way system. This reminds me of bad 1950s sci-fi movies: spaceship lands on Planet Earth, stretches out acquiring beam, dissolves unsuspecting victim into pool of particles, draws beam back inside the spaceship. That's what a proboscis is.

But wait, there's more.

The proboscis of each species differs technologically, depending on what the insect consumes. The butterfly's acquisition instrument is extremely fine-tuned. Sap-feeding butterflies have proboscises that differ from the proboscises of butterflies that feed on flower nectar and differ again from those that feed on blood.

The tip of a monarch's proboscis, used for feeding on flower nectar, looks fairly smooth. A question mark butterfly, with its folded wings that look like dead leaves, feeds on tree sap. The tip of the question mark's proboscis looks more like a mop and in fact acts like a mop.

The tip of the proboscis of the vampire moth, which sucks mammalian blood for a living, has sharp arrow-like projections that allow it to pierce flesh, including human flesh. Entomologist Jennifer Zaspel knows this firsthand. While she was collecting moths in Siberia one summer for her dissertation, one particular moth caught her eye. She captured it in a small glass vial. This fairly common moth, spread across much of Asia, was said to be a vampire, but no one had documented the behavior. For all Zaspel knew, the moth's reputation could have been completely undeserved.

She poked her finger into the vial.

The insect began probing her finger with its proboscis. Using the muscles in its head, the moth pierced her flesh and then began drilling down. The head moved back and forth in an oscillatory motion. Erect barbs on the proboscis tip acted like a saw, as the moth drilled deeper and deeper into her finger tissue.

"They pull it out and put it back," she told me. "It's almost like a sewing needle."

"Did it hurt?" I asked. I was incredulous. "Why did you do this?"

Having spent a good deal of time in Africa, I am instinctively suspicious of insects hoping to get inside my body. I definitely do not want them there. In my experience, such insects are up to no good.

"It doesn't feel good. It becomes painful after a while," she answered. "I don't know why I did it. I was just curious. That was really all."

"Will you do it again?"

"I don't know," she answered. "We'll see."

Her voice sounded dreamy when she said that, as though she was not entirely put off by the thought.

I told her about several famous scientists, including Charles Darwin, who had almost died by putting insects in their mouths.

"At least you're in good company," I said.

Zaspel suspects that the moth she captured does not survive on flesh, but on fruit. The insect needs the saw-like proboscis tip to penetrate the fruit's tough skin. Perhaps its propensity to saw into flesh is just a happy side effect.

A moth in Madagascar with barbs on the tip of its proboscis has recently been seen predating on the tears of sleeping birds. According to scientists, an "armoury of hooks, barbs and spines" allows the moth to penetrate the closed eyelids of the birds and to "anchor" the proboscis while it soaks up the tears. Insidious. The process seems not to harm the birds, who typically stay asleep, so that scientists speculate that the proboscis may be delivering some sort of compound, even a narcotic or antihistamine, that keeps the birds somnolent.

Tear-feeding is not limited to birds. In Thailand, there are moths that feed on human tears. "Humans experience pain" during the process, I read in one paper discussing this phenomenon.

I bet they do.

But why do butterflies and moths want tears in the first place? Or the blood of entomologists? Or even just tree sap? I had always thought that butterflies drank nectar and that was the end of the story. Wrong again.

The list of materials other than nectar on which Lepidoptera feed is stupefying, myth-busting, slightly nauseating, even frighteningly ghoulish:

dung, decaying plant material, bird droppings, fruit both fresh and rotted, crushed pollen, blood, decaying flesh, other Lepidoptera (preferably dead but not necessarily so), caterpillars, sap, human sweat, urine, beeswax, honey, fur.

Like us, these insects need "supplements" such as salts and proteins. This is particularly true for females, in that they must produce hardy eggs that can survive to engender the next generation. On the other hand, suggests lepidopterist David James, male butterflies in some species are more likely to do the seeking, and in a few butterflies, females don't do it at all. The caterpillar's task is to gobble up as many of these essentials as possible to store away for future use, but the adult flying insect will also need to gather its own nutrition.

There are, as usual, exceptions. The female cecropia moth that changed four-year-old Matthew Lehnert's future does not consume anything during her flying stage of life. Her only job is to reproduce and lay eggs, and as a consequence she lives only a week or so. Ergo, she has no proboscis. Why waste energy producing an organ that won't be used?

This habit of sitting on the ground and consuming something invisible to the human eye—"puddling"—was noted long ago, but it confused scientists. We assumed that butterflies and moths were somehow acquiring liquids this way, but since the insects were "puddling" where there were sometimes no puddles, the facts didn't add up. Then Lehnert and his colleagues Peter Adler and Konstantin Kornev began researching the proboscis with the latest generation of high-powered microscopes, which is how they discovered the curious perforations.

We owe this unexpected advance in knowledge in large part to the wanderings of two little girls who were chasing butterflies one lovely afternoon in the wilds of South Carolina. The materials engineer Konstantin Kornev was supervising his daughters in the great outdoors. When he noticed their attraction to butterflies, he helped them look more closely. Then he became curious himself: How is it that butterflies consume so many different foods? They seemed to drink water, to consume nectar, to enjoy a good meal of honey, a fluid that is inherently sticky and doesn't flow well. How do they do all these varied things? You can use a straw to sip water, or even sugary nectar. (Kornev was still thinking in terms of the conventional belief that the butterfly's proboscis was a straw.) But if you try to sip honey, you're not going to get far. Ditto for tree sap.

Kornev next asked himself a question that, incredibly, no scientist (even Charles Darwin) had ever asked: What—*exactly*—was going on? This is often how great science is done: by focusing on something so seemingly simple and straightforward and obvious that no one thinks about it. It helped, of course, that Kornev's specialty was the invention of new substances modeled on those found in nature. He was by training prone to thinking about natural materials on the micro level.

His curiosity might have remained merely idle had he not been presented with an entirely different kind of problem: what to do with a group of high school students who wanted to spend two summer weeks in his lab. They wanted to take on a science project. They hoped to complete the project by the end of the two weeks. And, they wanted to do science that no one else had ever done.

Tall order.

Kornev remembered the butterflies. Why not have the students film the process of what Kornev then thought would be "sipping" or "drinking"? You could put droplets of water containing various amounts of sugar on a counter and put a camera right beside the butterflies. You could then slow down the film and watch exactly what the butterflies were doing.

When Kornev and the students discovered that the proboscis tip was not, as was previously believed, inserted deeply into the drop of fluid, they searched the scientific literature for greater detail.

There was none. The "drinking straw" myth was the ruling paradigm. We had all just accepted the idea. Kornev teamed up with the biologist Peter Adler. They hired Matthew Lehnert as their graduate student. They began thinking much more deeply about evolution. How could an insect with a comparatively small body have enough stored energy to move liquid along the full length of its mammoth proboscis? It just didn't add up.

According to the physics of fluid transport, the hawkmoth with the huge proboscis predicted by Darwin should not be able to do what it clearly does. Imagine being asked to suck a liquid through a straw that was several times the length of your body. If you managed to do this at all, the energy it would require would likely be much more than the energy you acquired in the food you were trying to take in. Rather than a net gain, there would be a net loss. These are not good economics.

The team found that evolution has devised a solution: microdroplets. The liquid is moved up the proboscis in infinitesimally small beads of fluid, which are interspersed with bubbles of air. Transporting the liquid in these discrete "packages" substantially reduces the friction involved, which means that substantially less energy is required. The team is applying this innovative idea to human-produced fibers that, by mimicking nature's solution, can improve treatments in a variety of medical fields, such as gene transfer and wound healing.

Adler, Kornev, and Lehnert are also interested in Zaspel's vampire moth. How *precisely* did that moth consume Zaspel's blood? Blood is pretty sticky. If you step in a pool of blood at a murder scene and try to escape, it won't be hard to track you down. You'll leave prints, of course. But your footsteps will also make noise as the blood on the bottom of your foot begins to coagulate. It will be easy to hear you walk away from the crime.

How does the vampire moth keep from getting his proboscis stuck in the wound he creates? When the proboscis curls and uncurls, how does it keep from sticking together? Even more interesting: How do the tiny holes in the proboscis through which the liquid must flow not become clogged with coagulated blood?

"I'm interested in looking at the molecular properties of the saliva, to see if there are gene products that might be facilitating the movement of the blood," Zaspel explained. "Something is working very well for a species that is able to transport both nectar and blood. There's a lot to be learned. What structural modifications are externally and internally allowing this to work so well?"

There also might be certain molecules or compounds the vampire moth uses to keep the blood from gumming up the proboscis. If so, science wants to know what they are. This is much more than idle curiosity. Understanding how blood is transported through the miniscule tubing of the proboscis and learning about new, previously undiscovered anticoagulants could be the foundation for important breakthroughs in medical science and technology. For example, surgeons performing lengthy operations would be relieved of the pressure of coping with "sticky" blood.

All this diversity, some of which can be profoundly beneficial to humans, is the result of the spread of flowers. "Until you have the flowering plants," Lehnert told me, "proboscises were short and stub-like and fleshy, if you will, capable of feeding on exposed sugary fluids and water droplets." When flowering plants appeared, those proboscis-bearing flying insects blossomed into the glorious butterflies that keep us company in the modern world. In terms of evolutionary history, it happened in an instant in time.

This made me wonder: How much do we know about the ancient history of these insects?

Three

THE NUMBER ONE BUTTERFLY

One would hardly anticipate that creatures so delicate as butterflies could be preserved in a recognizable state in deposits of hardened mud and clay.

Samuel Hubbard Scudder, Frail Children of the Air

Just about 34 million years ago, on the east side of the rising Rockies, a river ran from north to south through a high valley. Along its banks stood groves of redwoods, many well over ten feet in diameter. The forest canopy towered 200 feet above.

In the shelter of this natural Notre Dame, butterflies fluttered. Painted Ladies, fabulous butterflies looking much like those common in our world today, enjoyed this primeval world. There were other Lepidoptera, too—many different species—and a variety of spiders, katydids, crickets, cockroaches, and termites and earwigs and water bugs. Tsetse flies twice the size of those found in modern Africa harassed the wildlife. Mammoth wasps predated upon other insects, no doubt swooping down upon Lepidoptera caterpillars. Bees also flew here. Indeed, this entire world greatly resembled parts of the world in which we live.

Mammals were plentiful. There were dog-size three-toed horses, and rhino-size brontotheres, long-extinct kin to horses. There were oreodonts, cloven-hoofed animals distantly related to our own modern pigs and deer. Birds, descendants of the dinosaurs who had disappeared tens of millions of years earlier, filled the air. Their loud screams mingled with the rustling of leaves. Opossums, then as now, grazed on the plentitude of insects.

All this vitality was supported by a wide-ranging cast of plant life. There were walnut and hickory trees. Palm trees and ferns and poplars and willows grew along the wetter areas of the banks of the river. The animals could browse sumac and currant bushes and on wild apple trees and bean pods. There were even cashews. Temperatures were roughly akin to those of San Francisco today.

It was not an easy life, though. A few miles distant from this seeming paradise stood violent volcanoes that erupted periodically, casting out turbulent flows of rock and mineral matter that swept down the mountainsides and piled up in the valley. Like cement, the liquefied debris encased the valley's living things, hardening around the bases of the redwoods. The debris rose as much as fifteen feet high around their impressive trunks, smothering the roots and killing the trees.

At some point during this period of tectonic violence, one eruption sent material down the mountain and across the stream. A dam appeared.

A shallow lake, lasting millions of years, covered the land.

Perhaps the butterfly was alive when it landed on the lake surface with its wings spread wide, as though it had been collected and pinned by sad old Herman Strecker. If so, why didn't it take off again? Was it caught up in a gust of wind and slammed down irrevocably? Did it struggle a bit against the surface tension of the water? Or was it perhaps caught in some sort of inescapable slime on an algal mat?

Whatever the reason, the perfect little butterfly slowly sank. Ash layers continued to build up, covering the butterfly. Cell by cell, the insect transformed into stone. Its features were so finely preserved that it's still possible, tens of millions of years later, to detect the butterfly's individual scales and to see some of the patterns on its once-lustrous wings. Modern technology may one day even be able to show us its true colors.

Many fossils other than this magnificent butterfly have been discovered in the lake where the ancient butterfly fell. The overtopping layers of ash and organic matter laid down season after season and year after year created what scientists call paper shales—ultrathin layers that are even today not particularly hard. If you carefully break apart some of these layers, you can still see the minutiae of the life preserved on the lake bottom—even the details of fish scales, and the bloodsucking apparatus of a tsetse fly, and holes in the leaves of plants where insects—caterpillars, perhaps?—have fed, and the fine particulars of the stem joints of a horsetail (tougher cousin to a fern), and pollens preserved in such detail that modern scientists can identify their species. But while fish and water bugs and plant leaves are common, a few of the finds are so rare as to be unparalleled.

The butterfly—now called *Prodryas persephone*—was one such discovery. Its detail is so precise that the woman who found it and the scientist who named it and the enthralled Victorians who flocked to see it were overwhelmed with its exquisite clarity. Even the delicate antennae are there, slightly curved to the left in death but still knobbed at the end, as a modern butterfly's would be. Some scientists have speculated that even its proboscis was preserved, but to find out it would be necessary to destroy the fossil to look beneath its head. Perhaps eventually some new technology will show us the coils just underneath its head.

Prodryas flew in a region we now call Colorado, in a place we now call the town of Florissant, just before the cataclysmic conclusion of a time we now call the Eocene—the Dawn Time, the time when, after the eradication of the dinosaurs, modern mammals first appeared on Earth, a time that began about 56 million years ago and lasted until just a bit less than 34 million years ago.

The butterfly's time was the last gasp of a period of exceptional warmth and exceptional rainfall on our planet, a petri-dish period when biological experiments of all kinds proliferated and various new forms of life emerged, when the earliest known horses and the earliest known true primates and so many other new mammals evolved.

It was also a time when flowers spread far and wide. Were butterflies spread worldwide, too, enjoying the flowers? It seems likely. We know that butterflies had been around long before the Eocene, most likely fluttering around the heads of Cretaceous dinosaurs, but did they get a jolt of encouragement during this tropical era? Maybe. We just don't have the evidence. Fossilized butterflies are exceptionally rare. The discovery of even a wing fragment is cause for celebration.

That's why Florissant is so important. The site has identifiable fossils of more butterfly species—possibly twelve—than any other site in the world. But none are as exquisite as *Prodryas*. Except for this perfect fossil, all we have worldwide are mostly fragments: bits of wings, scales, remnants encased in amber.

The Florissant butterfly is a singular jewel. When it was discovered, the entire world marveled.

Tens of millions of years passed after the butterfly landed on the lake surface. The world got colder. The world got warmer. The Pleistocene's ice ages came and went.

Some butterfly species evolved to adapt to the climate's changes, fluttering around for only a few weeks in summer when it was warm enough for their wings to spread and fly, and then burrowing down safely somewhere for the rest of the cold year. Other species moved their bases of operation to more comfortable regions.

The Florissant butterfly was apparently one of those. Were you to see the Florissant butterfly flying today, you would not find it flying in modern Colorado. Instead, it would inhabit warmer, wetter, tropical regions—regions that are today much like Florissant was then.

When the first people came to the Florissant valley, possibly 15,000 years ago, they would have been astonished by the redwoods-turned-to-stone. In Herman Strecker's time, Europeans looking for fossils came to the site. News of their marvelous discoveries appeared across the continent in distant cities like New York and Boston and London and Paris. Children's books with titles like *Wonders of the World* featured pictures of the trees. As a child, I had a much-read copy, given by an elderly aunt.

Scientists were captivated by the fossilized redwoods. In 1871 a collector from New York State's Cornell University, Theodore Mead, wandered by and collected several fossils to take back east. Scientists liked what they saw. News spread. Several years later, the first scientific expedition arrived. Others followed.

The area was "sacred ground for American paleontology," according to paleontologist Kirk Johnson. It was ridiculously easy to find fossils of plants, insects, and other tiny things. Train tracks were laid. Passengers in the late 1800s paid their fares for a day's adventure on trains with names like The Wildflower Excursion that ran from the town of Colorado Springs up to the valley. Fare-payers were allowed to explore for fossils and to carry away souvenir fossils, including pieces of the stone trees.

People took all sorts of things. A landowner used some of the larger chunks of fossilized trunks to build a hearth for a resort fireplace. One entrepreneur even tried to cut away a fossilized redwood stump in order to take it to the 1893 Chicago World's Fair. This endeavor failed because the saw blade stuck in the stone trunk. It is still there today.

One day the amusement park entrepreneur Walt Disney came by. Later he bought one of the fossilized redwood tree trunks. Weighing five tons and 7.5 feet in circumference, it's now in Disneyland, in California, near the Golden Horseshoe Saloon, an ice cream joint.

The early scientists and tourists and entrepreneurs did not discover the butterfly fossil. That honor lay with a woman, a settler who married at age thirteen and had seven children. Charlotte Coplen Hill was born in Indiana in 1849 and brought west by her family a few years later. She married in 1863. The couple homesteaded in Florissant in December 1874. By then Charlotte, twenty-five and the mother of many children, would soon become a grandmother and was mature beyond her years.

She realized the importance and value of what lay beneath her feet. The couple applied for a formal homestead claim in 1880. They ran cattle, grew crops, and built their ranch. Long before that, though, Charlotte had developed a curiosity about the ecosystem encased in the ancient lake bed. The fossilized redwoods would have been impossible to ignore, and it's likely that she often uncovered, perhaps working away casually at the end of a long day, the leaves and insects pressed in between the layers of paper shale. In any case, by the time the later scientific expeditions arrived in her valley, she had already set up a small paleontological museum. She had "boxes upon boxes full of fine paper like shales covered with the impressions of most perfect insects," researchers found.

It's easy to imagine them salivating. Her work was so good that as early as 1883 a fossil rose—*Rosa hilliae*—was named after her. Researchers depended on her heavily. At least one scientist, Leo Lesquereux, a pioneer of North American paleobotany, never came to Florissant, relying instead on Charlotte Hill to provide him with new plant fossils to describe. Harvard's Samuel Scudder, an avid lepidopterist as well as a paleontologist, visited the valley briefly, saw Charlotte's work, and realized he could simply buy what he wanted directly from her rather than doing his own fieldwork.

Scudder never publicly acknowledged her contribution to his science, much to the dismay of Florissant's modern-day paleontologist Herbert Meyer, one of Charlotte's biggest fans. Meyer describes her as self-made and deeply interested in her world. He suspects that Charlotte, a busy homesteader, enlisted her children in the search, encouraging them to look for buried treasure, just like children do today.

No one then had any idea of the fossils' true age. They knew that Florissant represented life "long ago," but science had yet to conceive of the several-billion-year history of our planet. As late as 1908, paleontologist Theodore Cockerell published an enthusiastic description of Charlotte's land, eloquently likening it to Pompeii: "In ancient times—say about a million years ago—the valley was the site of a beautiful lake—Lake Florissant. This body of water was perhaps about nine miles long, but narrow, and strongly indented by wooded headlands at every point. Here and there were small islands, upon which grew tall redwood trees and other vegetation. It was just such a place as would have delighted the heart of Fenimore Cooper and his hero of the Leatherstocking tales."

Today, one million years seems like only yesterday. If you had suggested to him then that the fifteen-square-mile lake bed was 34 million years old, he would not have believed you. Few then could imagine such an expanse of time.

• • •

When Harvard's Samuel Scudder, paleontologist and avid lover of living butterflies, received the fossil, he knew he had a stupendous prize. The butterfly was "so perfect as to allow a description of the scales," he gushed, adding that it was the first such found in America. In 1889 he published *The Fossil Butterflies of Florissant*, and a decade later a children's book, *Frail Children of the Air*, which discussed the Florissant fossils. At least one young reader, Frank Carpenter, chose a career in paleontology after encountering this book: "And there was a picture of this fossil butterfly from the Florissant shales of Colorado, with its wings outstretched and all the color markings. When I saw it, my eyes bulged. I told my father when he came home from work that what I wanted to do was work on fossil insects." Carpenter went on to become one of North America's leading experts on paleoentomology.

Born in Boston in 1837, Scudder did not become a butterfly fanatic until sixteen when he attended Williams College, in the hills of western Massachusetts. His father was a middlingly successful businessman and his older brother a missionary. He was by no means slated to become a scientist, but butterflies changed that. At Williams he met another student who showed him a box of twenty or so pinned butterflies collected nearby. I can only imagine how furiously the neurons in his brain's visual pathways must have fired, seeing colors like that for the first time.

"I had not dreamed," Scudder later wrote, "that such beautiful objects existed, least of all at home, or that so many different kinds could be found in one spot." Immediately after seeing that box, he began to collect his own specimens. Once, when he caught a particularly rare and beautiful butterfly, he was so overwhelmed that he resorted to citing lines from Shakespeare.

From Williams College Scudder went to Harvard to study under biologist Louis Agassiz, whose antievolutionist ideas he thoroughly absorbed. Eventually, he founded the American branch of insect paleontology. Scudder named Charlotte Hill's butterfly *Prodryas persephone*, giving it its own genus and species name. He placed it in its own wood-framed box and gave it the catalogue number "1." He was so proud of it that he took it to London in 1893 to exhibit at the Royal Entomological Society.

Strangely, though, at one point he seems to have tried to sell the fossil for $250. In a publication called *The Canadian Entomologist*, in 1887, on page 120, an advertisement appeared: "Fossil Butterfly for Sale." The text read: "In order to illustrate more fully his forthcoming work on New England Butterflies, the undersigned offers for sale for Two Hundred and Fifty Dollars, that wonderfully preserved Fossil Butterfly, *Prodryas Persephone* [*sic*], of Colorado. . . . Less than twenty specimens of fossil butterflies are known in the world, and this is by far the most perfect and best preserved."

This offer is shrouded in mystery. Why would he sell such a treasure? No one knows. Perhaps, suggests Herbert Meyer, it was only part of the fossil. Maybe he needed money. Today records would be kept about such things, but during the Victorian era paleontological details were not meticulous.

Whatever the reason for the proposed sale, it did not happen. Scudder's fossil remains intact and safe in a basement vault in Cambridge, Massachusetts, at Harvard, more than two thousand miles from where it once lived. It is still catalogued as Number One.

Almost 150 years after Charlotte Hill unburied *Prodryas*, I went to Harvard to pay my respects.

This esteemed butterfly that touched the souls of so many humans is now safely encased in the cleanest fossil storage area I have ever seen. Most fossil museums and storage areas are dusty old things. They remind me of my grandmother's attic. But Harvard's was brand-spanking-new, so pristine as to be completely dust-free. Just to check, I ran my forefinger over a surface when no one was looking. I was right. The place was hospital-clean.

My guide was Ricardo Pérez–de la Fuente, an inestimably gracious scientist from Barcelona. We walked down the basement hall and through a newly installed glass door, then down a long line of locked cabinets. At the beginning of that line, in the first cabinet, was the top-billed Number One Butterfly.

All the butterflies from Florissant that were brought to Harvard by Scudder are honored in this way. The Number Two fossil is the next most important Florissant butterfly, and so on, down the line.

I am by nature not prone to iconization. I was prepared to be skeptical. But when Pérez–de la Fuente unlocked the cabinet and brought out the specimen, I surprised myself. My emotional response was suitably reverential.

Prodryas is still encased in the wooden glass-topped box made long ago to hold it. We handled it like a relic. I admired it through the glass. Then we carried it to the next room.

I made a tasteless joke about dropping it and received a polite laugh that made clear that I shouldn't be such an idiot.

We looked at the fossil under the microscope. It was possible to see the scales. The veins running through its wings were quite clear. (These are not veins in the human sense, but structures that transport oxygen and firm up a butterfly's wings.) We looked at tiny, whisker-like filaments extending from the head.

I loved seeing the scratch marks in the stone surrounding the butterfly, made by the artist who had so slowly and patiently uncovered the fossil from its gossamer-thin covering. Preparing this fossil must have been exciting but also terrifying. The artist—Charlotte Hill?— must have had to work with near-surgical precision to remove the material covering the treasure beneath. Had her hands been too heavy, she would have removed parts of the butterfly itself. The butterfly had a small tail on each wing, not as long at the tails of some of today's swallowtail butterflies, but easily visible. One of those tails remains intact. The tip of the other is gone. Was it a mistake made by someone who handled it?

I remarked on the steadiness of the hands that worked on this gem.

"There are two professions in the world," Pérez–de la Fuente answered, "that must have these steady hands. Neurosurgeons and entomologists."

"What colors did it have?" I asked again, expressing frustration and longing to see it alive in its ancient world.

"In paleontology," he answered, "we are full of uncertainty. And the beautiful part of this is that we embrace the uncertainty."

Later he added: "This butterfly had a great impact on the field. It is one jewel that led to the discovery of more jewels. That's life. That's how science works. It's a beautiful concept."

Nearly one-third of the world's described fossil butterflies come from Florissant. At least twelve different species have been named to date. One scientist has called the lake bed an "insectan Pompeii."

Yet the site was almost lost to a 1960s housing development. The Florissant valley was a perfect vacation spot. There are great places for hunting and fishing, verdant hiking areas, trails for horseback riding, lakes for swimming and boating. In the valley's southern end, A-frame houses on tiny lots appeared. Land speculators sniffed around.

Meanwhile, in 1959 the National Park Service began to study the area to see if it should become a national monument. The report, issued in the early 1960s, recommended protecting the fossil beds. This had been discussed since the early 1900s, but nothing had ever happened. Now the appearance of the tiny houses on the tiny lots created alarm. Paleobotanist Harry MacGinitie told federal elected officials that although the land was not much good for farming, "as a page of earth history from the dim past it is priceless. . . . There isn't anything else like it."

The valley had become an "economic temptation," one advocate worried. Conservationists and scientists became involved, including paleobotanist Estella Leopold, daughter of the renowned author of *A Sand County Almanac*, Aldo Leopold. Federal land-protection bills were mired in committees.

Land-protection advocates were eloquent: building houses on the fossil beds would be "geologic book burning," like "wrapping fish in the Dead Sea Scrolls" or "using the Rosetta Stone for grinding corn."

The phrases stuck. National newspapers covered the story. The great political cartoonist Pat Oliphant drew a bulldozer-driving developer who looked like Snidely Whiplash and an environmental advocate with the muscles of Paul Bunyan. In a *New York Times* story on July 20,

1969, "Fossil Beds in U.S. Go Unprotected—Government Fails to Act on Florissant Purchase," one scientist said, "There is only one irreplaceable volume on this subject in the universe." A *Denver Post* headline that summer ran: "Florissant Project Still Petrified."

Nevertheless, by summer's end, bulldozers were at the ready. Women and children showed up with picnic baskets and bedrolls. Their plan was to surround the machines with live bodies.

Oddly, the 'dozers never moved. The drivers were delayed at a bar down the road, where free drinks may mysteriously have been made available. To this day, no one knows how this happened. Perhaps the men did not want to face the women and children. People who knew people who knew other people may have convinced the men to have another beer.

And just as the speculators were preparing to move ahead anyway, the federal decision appeared: Florissant would be a national monument, off-limits to developers. Republican president Richard Nixon, an advocate for the environment and in fact the creator of the Environmental Protection Agency, signed the bill creating Florissant Fossil Beds National Monument on August 20, 1969.

Other locales hold similar fossils. Nearby is a privately owned parcel open to the public. At the Florissant Fossil Quarry, small slabs of shale are cut out of the hillside and brought to picnic tables. For ten dollars an hour, kids can pile into them and separate the pages to peer at the life that might be inside.

On rare occasions, something truly fabulous shows up. Anything of real value has to be turned over to authorities, but the kids then get their name attached to the specimen as the discoverer. School groups come often. One child opened up one of these slabs and inside found an entire fossilized bird.

Elsewhere in Colorado, farther to the north and west, are fossils that are even older. The Green River fossils, about 50 million years old, are, like Florissant's, preserved within layers of shale. But that's where the similarity ends. These fossils, while unmistakably butterflies, are more fragmentary, and are found throughout an extensive shallow lake system that spread, like today's Great Lakes, across millions of acres. There are Green River fossil sites in Utah, Wyoming, and Colorado.

Because of these, we know that butterflies were not unusual 50 million years ago. And because of fossils encased in amber in Denmark that date back even further, to 56 million years, we know that butterflies existed then. But no one knows for certain how old butterflies truly are.

Paleontologists widely accept that the first flowers, about 140 million years old, drove butterfly evolution. Conrad Labandeira, an expert on the butterfly-flower partnership, suggests that butterflies might not have been widespread until well after the earliest flowers. "The first flowers were bowl-shaped," he explained to me. "They did not require long proboscises." Slowly, he suggests, over millions of years, the pairing between flowers and butterfly proboscises became more specialized—until those pairings reached the extreme noted by Darwin.

By the time flowers first evolved, moths had already existed for at least 50 million years. Labandeira and several colleagues have found fossil evidence of early moths in a rock layer in China dating back 160 million years. Those moths had already evolved primitive proboscises, which they used to feed on sweet pollination drops produced by gymnosperms, including conifers such as pine trees, cypresses, and redwoods. As anyone with one of these trees knows, in spring they release extreme (some might say excessive) amounts of pollen (in my yard, the pollen is the color of bile) that effectively coat everything and anything that's nearby. (This includes my bright-red Prius.)

It costs a lot in energy to produce these copious amounts of mostly wasted pollen. Flowers were a much-improved reproduction strategy, particularly as flowers became more sophisticated over tens of millions of years. So, too, did the pollinating insects.

The precise flower-insect coupling is now called "faithful pollination." The implied association with marriage is not accidental. When a species of flower can seduce a certain species of butterfly or moth, then that flower can successfully reproduce much more cheaply, energy-wise.

The seduction of butterflies by flowers is nothing new. Labandeira pointed out that over the course of evolution, since the days of the first insects more than 400 million years ago, plants had seduced a wide variety of insects into developing long proboscises to serve the plant's needs at least *thirteen different times*. Far from these relationships being adversarial, Labandeira suggests, they may well be consensual.

"Many of the relationships between butterflies and their plant hosts represent what was once thought to be an antagonism, but which now turns out to be really mutualism," he told me. You do for me and I'll do for you.

So it turns out that Darwin's orchid-moth match was no mere accident. Rather, it was a strategy perpetrated by plants on innocent insects over and over and over again.

After all, flowers are not always benign rulers. Some orchids engage in cruel trickery to entice bees to enter their lairs. It turns out that the most obscene-looking orchids *intend* to look obscene. They send out certain visually sensual clues to male bees, which respond by flying to them and exhibiting a certain graphic mating behavior that I won't delve into, except to say that when the males are done, they fly away apparently satisfied—and covered with orchid pollen.

At the fossil lab at Yale's Peabody Museum of Natural History, Susan Butts and I looked at some of the butterfly fossils preserved in amber. We had already had our discussion of what makes a butterfly a butterfly and what makes a moth a moth, and we had moved on to look at fossils from the museum's lavish Green River insect collection.

Then Butts pulled out the amber collection. Amber, hardened and aged tree resin, has been a material revered by humans for thousands of years. A somewhat rare material, it is harvested in copious quantities at only about twenty sites worldwide. Some of these sites have long histories. Ice Age people carved figures of horses and other animals out of amber, just as they did out of ivory and antlers. A 4,000-year-old stylized horse has been found in Poland, and amber artifacts have been found at Britain's Stonehenge. Chinese artisans have been carving intricate amber sculptures for thousands of years.

To paleontologists, though, amber has an entirely different value: as a preservative that can show the organism in three dimensions. As tree resin flows down a trunk, it can encase whatever may be in its path, from leaves to seeds to pollen to insects. Remarkable objects have come down to us this way. An inch-long fossilized three-dimensional tree cone from 140 million years ago, found in what's now Kazakhstan, gives us a clue to what the world looked like at the height of the reign of dinosaurs, just as flowering plants took over the planet. The conifer cone is easily recognizable as such to anyone today.

By the end of the reign of the dinosaurs, flowering trees would cover many locales where cycads and conifers once dominated. In a few places, resins from these new trees would preserve entire ecosystems. In the Dominican Republic, prolific amber mines have revealed a plethora of life-forms, including myriads of beetles, delicate damselflies, insects called treehoppers, termite swarms, fly swarms, and several kinds of moths and butterflies from roughly 25 million years ago. Susan Butts once had a wedding ring of Dominican amber with an insect inside, bought when she and her now husband honeymooned there. Sadly, amber is vulnerable. "Amber jewelry is a bad choice for geologists," Butts said. While she was using her geology tools, the ring broke. She now has a less vulnerable ring made of platinum.

We looked under a microscope at an amber-encased butterfly from Tanzania, dated to about 4 million years ago—about the time when early proto-humans walked the plains there, and encountered three-toed *Hyperion* horses.

"You can see the eyes," Butts pointed out as we looked at the insect, "the antennae where they attach to the head, where the head attaches to the thorax. You can see the legs. See this thing? See the round circle? That's its coiled proboscis. That's it, coiled up right there."

I counted the one-two-three-four coils that were easily visible, frozen in time. It was as though we were looking through a crystal ball, able to peer directly into a long-gone world. The butterfly was easily recognizable as a butterfly, not much different from what we see today. Thinking about that, 4 million years didn't seem to be that long ago.

In the Yale collections, as elsewhere, fossil butterflies—even fragments—are rarities. Of the roughly 17,000 insect fossils the museum owns, only 61 are Lepidoptera, and most of those are not identifiable as either moth or butterfly. Butts also pulled out the museum's few specimens of butterfly fossils from the Green River Formation, collected from the northwest corner of Colorado by Yale volunteer Jim Barkley, a retired geologist. They were fragments, none as perfect as the Number One Butterfly, but they were so much older, almost (but not quite) the world's oldest known butterfly fossils. (The oldest currently known are from about 56 million years ago, encased in amber from the Baltic Sea.)

Barkley owns his own personal fossil site where he prospects routinely in all but the wintriest of wintry weather, when it often gets down to 10 or 20 degrees Fahrenheit below zero. He sends all of his insect finds to Yale. Of the more than 6,000 specimens he has donated so far, only a few are of butterflies.

"Maybe there are others, not yet sent?" I asked.

Butts dubbed our proposed venture the Yale-Barkley-Williams Expedition. Sounded good to me.

We all met up—seven of us, ranging in age from 6 to 66—early on July 1, at a scenic public park along the Colorado River. The river was beautiful, but at 9:00 a.m. the day was already roasting hot. We piled into our vehicles and climbed north along Highway 13, up toward the much cooler Roan Plateau.

Only a short way up the highway, at an entirely unprepossessing site, Barkley pulled over. Acres of tumbled-down shale awaited us, just beneath towering cliffs of yet more shale. And more shale.

An earlier prospector had brought a large yellow machine and simply knocked down the cliffs. The shale had cracked into tiny pieces, and it was our job to sit on the burning rocks, sift through the fragments, and open up the layers to see what—if anything—was inside. I thought of those old Technicolor movies about ancient Rome, the ones where luckless captives toiled endlessly on hot Roman hillsides.

This looked pretty discouraging.

Facing this dismal task, Barkley, Butts, and her colleague Gwen Antell, who was about to head off to Oxford, looked as pleased as punch.

Remarkable fossils began appearing. Even *I* found an insect.

I said I was impressed by the cornucopia of fossils, but admitted I wished we'd found a butterfly.

Jim Barkley looked inexplicably pleased.

"Time to go home," he said.

Barkley's small homestead ranch house is given over not quite but *almost* entirely (he has a wife) to his love of paleontology. He has a high-tech microscope on a boom stand with a 10-megapixel camera attached. For each specimen, he takes five to twenty pictures at varying foci, then uses stacking software to meld the photos into a single focused (hopefully) frame.

There were wires everywhere, lots and lots of bottles of water half finished on worktables, family pictures, guidebooks, reference books, religious books, and, of course, stone dust.

"After dinner," he said, "I'll show you something."

People arrived to discuss the day's discoveries, so we all sat squeezed in at a small table, consuming grilled chicken and salad and beer. When the party broke up, Barkley and I returned to the work shed.

He pulled open a drawer.

Inside lay a nearly intact Lepidoptera wing, showing not only the veins in the wing, but some of the markings.

Why, I wondered, was a 50-million-year-old insect so easily recognizable to a layperson today? The first time I saw a 50-million-year-old horse fossil, I took it for a dog or a cat. Mammals have evolved considerably over time, but, apparently, not butterflies?

"That's because insects are perfect," Gwen Antell answered with a smile. They don't need to evolve.

"Arthropods were the first things on land, and three out of four animals today are insects. They have ruled the Earth for hundreds of millions of years. What's left to perfect?"

She was joking, of course.

Sort of.

The next day I drove on to pay homage to the former gravesite of the Number One Butterfly fossil. Florissant National Monument now has an extensive visitor center that explains both the area's deep-time and recent history.

The first thing I noticed in the visitor center was a sign: "Science is an ongoing process, not just facts." That explains, in a nutshell, why evolution is still called a theory: Not because it's incorrect, but because our understanding remains incomplete. Our knowledge of the hows and whys of change is, like change itself, always developing, always improving.

In the visitor center was an artist's interpretation of *Prodryas*. In this image, the wings were reddish with three black dots near the outer edges of the forewing and with white sections on both the forewing and the hind wing. Three smaller black dots enhanced the edges of both hind wings.

The proud proper Bostonian Samuel Scudder was quoted—"the finest butterfly fossil found in America"—along with the information that the easily visible wing veins show *Prodryas* to belong to a large family of butterflies known as the Nymphalidae—as do today's monarch butterflies.

On the wall was an 1878 map showing the ancient, now long-gone lake. The site of the petrified Big Stump was pointed out with an arrow, as was "Mr. Hill's" place. (Although Charlotte Hill apparently made the discoveries, she was not acknowledged. Perhaps she was back in the pantry when the men visited.) In the back room of the visitor center, in a heavily protected cabinet, was a fossil of a nearly perfect insect that looked remarkably like a modern bee.

In a place of honor is a photograph of a birthday cake celebrating Charlotte Hill's 160th birthday. In the center of the cake, done in rather good detail, is the outline of *Prodryas*. The monument's director, Herbert Meyer, who is determined that Charlotte Hill finally get the credit she deserves, invited her descendants to a birthday celebration for their ancestor. Many had never heard of the pioneer woman, the thirteen-year-old bride.

After we talked, Meyer and I walked through the monument's out-door areas. We visited what little is left of the fossilized redwood trees and ended our talk at the site of one fossilized stump out of which a new conifer was growing.

"As it should be," I remarked.

"As it should be," he agreed.

We were pleased: one life-form growing out of another. Evolution is about change, but it's also about continuity.

Four

FLASH AND DAZZLE

The wings of a butterfly are the only place where the laws of evolution are printed in color on a single page.

G. Evelyn Hutchinson

Charles Darwin was certainly not the only person to conceive of the importance of partnerships among our planet's plant and animal species. Indeed, that concept—now called ecology—was first discovered not by Darwin, or by any other famous Victorian man, but by a seventeenth-century teenage girl.

Maria Sibylla Merian would become famous as a lover of Lepidoptera, admired for her bravery, worshipped for her artistic finesse, revered for her scientific rigor—just as Charles Darwin is today. Then she would be forgotten, buried in the mists of time. She lived in Europe during the 1600s, an extremely precarious time for women. Both Merian and Hill entered into their adult worlds at the age of thirteen, but whereas Charlotte Hill began raising a family then, Maria Sibylla Merian at that age began her lifelong study of caterpillars, butterflies, and moths—and the plants that made their existence possible.

Merian's era ranks among the most astonishing in human history. It was a bizarre age—nightmarish and surreal, but also progressive and futuristic, highly technological, and downright exhilarating. For some, life in Europe was terrifying. For others, the culture was dynamic and enthralling.

The Thirty Years' War, waged across much of the continent on both religious and nationalist fronts, ended the lives of 8 million Europeans, and profoundly destabilized the continent. At the same time, though, new technologies and a burst in international trade created disposable income for the expanding middle class for the first time in history. When education opened to the masses, the response was overwhelming. Public lectures in the sciences were often standing room only. Even women were allowed to attend.

The century did not start this way. In 1600, the mathematician Giordano Bruno was burned in Rome because he insisted that the Earth and other planets circled the sun. The Jesuit priest Martin del Rio's book *Investigations into Magic*, a best seller published in 1600, fanned the flames of mob lunacy and witch burnings, resulting in 50,000 executions over the century. It was a dangerous age for women. Most of the victims of this mass malevolence were female.

Still, the Age of Reason gained ground. Pioneering technology allowed humanity to see the world in a novel—i.e., fact-based—way. At the forefront of this revolution were glass lenses that allowed people to peer into the world of the infinitesimal. Lots of people owned these "flea glasses," handheld instruments that allowed close-up study of insects.

People could look into a drop of water to see the hitherto invisible single-celled life-forms, revealing worlds within worlds. Until then, we humans had had no idea that such "animalcules" as amoebas existed. Repercussions ensued.

In a major cultural shift, knowledge became fashionable. Europe in 1600 had personified an Era of Unquestioned Certitude: God had ordained a rigid hierarchy. If you were born into poverty, that's what God wanted for you; remain obedient, starve without malice, and wait for your just rewards in Heaven. Striving for upward mobility was a sin.

Kings were divine. Everyone knew this, so there was no need to prove it. (Queens? Well . . . not so much.) The *scala naturae*, the "ladder of nature," literally, or the "natural order" of Aristotle's Great Chain of Being, classified all living things, from the "lowest" to the "highest." Every single being on the planet was either inferior to or superior to every other living thing.

This pecking order was encyclopedically detailed. For example, birds ranked below mammals. Within that category, birds of prey reigned over birds that ate carrion, which were superior to birds that ate worms, which were superior to birds that ate insects. Dogs enjoyed a rather high status, but not quite as high as lions, which after all were wild and free and powerful—and dangerous to boot. Only a short step above lions were women, who of course ranked well below men, who ranked just below angels, who ranked below God.

Insects ranked quite low on the ladder, just above plants and corals.

Except for butterflies.

Butterflies were special. Highly revered, they occupied their own private rung on the ladder, well above other insects. They were granted this privilege in part because of their irresistible flash-and-dazzle beauty and in part because of their mystery. They appeared to emerge spontaneously from hidden places and to fly off to the heavens. They seemed blessed by God.

Caterpillars, on the other hand, were worms—abominations worthy of the utmost contempt. Their place on the ladder of nature was very, very low. Slimy. Repulsive. Primeval. Consider Shakespeare's writings: Lawyers, despised by the Bard, were called "false caterpillars." Political advisors, whom he disliked, he called "caterpillars of the commonwealth," because they were eating their way through England's green and leafy land.

To understand this butterflies-versus-caterpillars discrimination, it's important to understand that Europeans believed that caterpillars and butterflies were entirely unrelated entities. This was accepted wisdom. Incredible as it may seem to us today, people did not connect a specific type of chrysalis made by a specific caterpillar with the specific type of butterfly that later emerged.

"If they did connect a larva with an associated adult, then it was supposed that some fantastical transformation had taken place such that a whole new animal appeared," explains entomologist Michael Engel.

This false belief existed because no one had actually *studied* caterpillars and butterflies. While they did understand the life cycle of the silk moth—they had been manufacturing silk for centuries—people failed to extrapolate to all Lepidoptera.

To people in 1600, the magic of the butterfly was that it emerged from what seemed to some to be the disgusting goo of a chrysalis. Figuring out that the emergence of a butterfly was an orderly process moving from egg to caterpillar to chrysalis to a gorgeous flying insect—and that this included *all* butterflies—was, writes author Matthew Cobb, "a remarkably complex challenge."

Discovering the truth about the egg-to-butterfly transformation helped science free European civilization from its *scala naturae* cultural straitjacket. In its place, the concept of an interdependent *net* of living things began to emerge.

It was Maria Sibylla Merian who accomplished this.

If you cut into a butterfly chrysalis, an abominable and noxious fluid oozes forth. Or at least it seemed abominable and noxious to people then. And yet, wait long enough and a dazzling butterfly, incandescent and flashy, slowly pushes its way out of its enclosure. In 1600, this was clear evidence of some kind of abracadabra. Sorcery or bewitchery or just plain good fun in the netherworld.

Beliefs that butterflies are magical is an idea that's as old as humanity. The Greek word *psyche* carried the dual meaning of butterfly and soul. Early Greeks believed that just as the butterfly emerged from its "tomb" to flutter off mysteriously to parts unknown, so did the human soul shed its earthly bonds to flutter unto the heavens.

On the other hand, caterpillars were "Devil's worms." A 1624 meditation by John Donne excoriated "serpents, and vipers, malignant, and venomous creatures, and worms, and caterpillars, that endeavour to devour" the world.

The misunderstanding was due to a firm belief in spontaneous generation, which everyone knew was a fact of life. Maggots spontaneously emerged from meat. Place dirty underwear and wheat in a glass jar and mice spontaneously generated. "Even learned people accepted that in some cases a woman could give birth to a rabbit or a kitten," writes Matthew Cobb. Shakespeare wrote that crocodiles spontaneously generated from Nile River mud. Bees magically appeared from the carcasses of rotting oxen. The century's own Johannes Kepler, smart enough to work out the mathematics of our planet's elliptical orbit around the Sun, wrote that caterpillars spontaneously generated from the sweat of trees.

Belief that the world was mercurial had consequences. In *The Astronomer and the Witch*, Ulinka Rublack tells how Kepler himself had to postpone his research to keep his own mother—charged with riding a calf backwards to its death and of turning herself into a cat—from being executed.

I asked Rublack if Kepler believed in witchcraft.

"There is nothing to tell us that he did not believe in witches, like pretty much everyone at the time," she answered.

No one was safe. Everyone was suspect.

Into this world of magical thinking was born the supremely rational and artistically talented Maria Sibylla Merian, a woman whose clarity, diligence, and persistence was almost itself a spontaneously generated miracle. Uneducated, relegated to housekeeping and cooking, she would, due to her love of butterflies, nevertheless establish a new standard in natural history and epitomize the century's most important scientific innovation: careful observation. When Merian was born in Frankfurt am Main (Germany) in 1647, looking carefully for facts was unusual. For all practical purposes, there was no such thing as "the scientific method."

At the age of thirteen she fell in love with caterpillars, despite their bad rap. Far from despising them, she thought they were beautiful. She discovered that many were quite loyal, eating only certain plant species while avoiding others. She followed those caterpillars as they emerged from eggs, then shed their skins several times while growing, then became chrysalises. She noted that each caterpillar, emerging from its chrysalis, became a specific butterfly.

Her groundbreaking discovery changed the world of science. Merian was more than just a careful notetaker. She drew and painted what she saw, providing much-needed visual evidence of her discoveries in these days before photography. These watercolor depictions, visually stunning, provided scientific evidence of what she observed. By studying caterpillars, butterflies, and moths for more than fifty years, she persevered to provide a considerable body of substantive evidence showing that spontaneous generation was pure nonsense. Instead, she showed, the natural world was ordered and rational, and partnerships within that world were consistent and dependable. Relationships were not haphazard.

We still have her research available to us today. Denied the right to publish in conventional journals because she was a woman, she self-published her research—both her notes and her artworks—in Germany and later in Amsterdam. Because of their scientific precision and artistic beauty, these became immediate best sellers.

Denied research funding, too, because of her gender, she would nevertheless undertake, with her own money and with only her daughter for company, Europe's first single-purpose scientific expedition to the Western Hemisphere. Unique as she was, she would come to be revered in scientific circles before her death in 1717.

All this—for the love of butterflies.

To say that her behavior was aberrant for a seventeenth-century woman would be an understatement. By walking through fields and gardens picking up caterpillars, she risked being called deviant, or perhaps even a witch. Lest you think that I am exaggerating her peril, recall that in *The Voyage of the Beagle*, Darwin wrote about a German scientist on South America's west coast who was jailed for necromancy because he took caterpillars into his house, which then became butterflies. And this was in the nineteenth century—two hundred years later.

Yet Merian got away with it. As far as we know, she was never accused of witchcraft or threatened with a trial for unwomanly behavior. Instead, she was likened to "the goddess Minerva," and her work was called *vervunderns*—astonishing. She was praised for her "tireless diligence." The "father of French entomology," René-Antoine Ferchault de Réaumur, praised her for her "amour véritablement héroïque pour les insectes"—her truly heroic love of insects. On the day of her death, Peter the Great bought up most of her work in her Amsterdam studio.

Merian's research has echoed down through the ages. Linnaeus used her books to help establish the field of taxonomy. In the nineteenth century, the great lepidopterist Henry Walter Bates in his classic *The Naturalist on the River Amazons* affirmed one of her findings that earlier researchers had ridiculed. American Victorian lepidopterists like Samuel Scudder would sing her praises. In the twentieth century, in *Speak, Memory*, the renowned novelist and lepidopterist Vladimir Nabokov named her as an important childhood influence.

Recently, art historian Gauvin Alexander Bailey called her "one of the most extraordinary figures in the history of science." Naturalist David Attenborough featured her work in *Amazing Rare Things*, published by Yale University Press in 2007. In 1995, in *Women on the Margins*, historian Natalie Zemon Davis wrote that she was "a pioneer," the first ecologist: "curious, willful, self-concealing, versatile, carried along through religious and family change by her ardent pursuit of nature's connections and beauty." Biologist Kay Etheridge calls her work "unprecedented," adding that Merian made a "seminal contribution" that "established a new standard in natural history."

In 2014, almost 300 years after her death, the Maria Sibylla Merian Society was formed, and in 2016 an exquisite facsimile of one of her books was published by a Dutch museum, more than 300 years after the original publication date.

I bought one. Turning its pages, I was awestruck by its beauty and detail. Yet I had never heard of her.

I wanted to know more.

Merian was not born wealthy, but her family did enjoy other advantages. They were in the artist/printer/publisher business in Frankfurt. Frankfurt was a cool place to be, a self-governing free city on the edge of change, a key center for books and the intelligentsia. Merian began learning the publishing arts from an early age. By the time she was born, the city's now famous book fair was already more than a century old. Books would have been a way of life for her.

Forbidden to use oil paints—these were for men only—she began by painting flowers using watercolors, and became an expert at mixing pigments to match the colors she saw in nature. Few artists then worried about showing insects with such exactitude, but she did. She wanted to echo the world's beauty as faithfully as possible. She wanted her red pigment to match the precise red of a flower petal or a butterfly's scales or of the colors of a caterpillar.

In her father's workshop she helped draw and hand-paint catalogues advertising flowers for sale. This was the century, after all, of Tulip Mania. Selling flowers was big business, and her illustrations had to be accurate for customers to know what they were buying. No doubt she helped produce the scientific works her father published. She most likely read some of these books and may have heard some of them discussed. Perhaps she met authors in her family's workshop.

The scientific debate of the century was over the genesis of living things: From whence came life? If not spontaneous generation, then what? A few scientists suggested that life—*all* life, even human life—derived from *eggs*, as with chickens. Neither magic nor alchemy was involved. This claim was as much a stab at the established order as Charles Darwin's theory of evolution would be two hundred years later.

Entering the fray, Merian began keeping formal notebooks that recorded in field notes and watercolors her research into the lives of caterpillars, butterflies, moths, and their favorite plants. For fifty years, she studied what they ate, how they mated, what their chrysalises looked like, and what their eggs looked like.

Merian became the world's foremost expert on lepidopteran life cycles. Others studied individual species, but no one understood the entire system as she did. This thorough knowledge allowed her to prove that (1) butterflies mated, (2) eggs were laid, (3) specific caterpillars hatched from the eggs, (4) those caterpillars consumed specific plants,

(5) after a predictable period of time, caterpillars created chrysalises, and (6) after a predictable period of time, specific butterflies emerged.

For us, this is shoulder-shrugging stuff, but for people in the 1600s, depicting a dependable life cycle was groundbreaking: there was method to the madness of living. Merian's renderings of caterpillars were unsurpassed. She must have had the steadiest of hands: in some illustrations, each hair on each segment of a caterpillar is meticulously drawn. Aided by her day's magnifying glasses and a primitive "microscope," Merian was able to look at the exquisite detail of caterpillars for the first time.

Research means nothing if you don't publish. And so at the age of thirty-two, married and with two children, she decided to publish her own book. Entitled *The Wonderful Transformation and Strange Floral Food of Caterpillars* (there's actually a much longer complete title, but you, dear reader, won't feel like reading it), the 1679 book was a smash hit. Readers clamored for copies.

"I think it necessary," she wrote in her introduction, "to state clearly that in general, all caterpillars, as long as the insects have mated beforehand, emerge from their eggs." She presented copious evidence. Her first caterpillar book had 50 examples, as did her second and third books—150 in all.

Not many copies of the original book exist today, but New York's American Museum of Natural History has one. Research librarian Mai Reitmeyer kindly showed it to me.

In general, these small and fragile tomes have not weathered well. To show me the museum's copy, Reitmeyer pulled on surgical gloves and suggested that she turn the pages while I looked.

Both of us leaned over the work to peer at the miracles on each page. We marveled at both the explicitness and the exquisiteness of the engravings.

Details of each plant were painstakingly rendered. Leaves, when shown as having been eaten by caterpillars, had exactly the kind of leaf damage that particular caterpillar would have made. In some illustrations, leaves were completely denuded except for the leaf veins—which was again what that particular caterpillar species would have done on that particular plant species.

Her depictions of a caterpillar's *instars* (various stages of larval development) were superb. Merian took the time to record each and every spot on the animals' bodies and matched their true colors closely because of her outstanding knowledge of pigments. She dotted one green caterpillar with the same miniscule rows of golden-yellow dots that she saw in the living specimen.

Even more remarkable was that she showed these various organisms *in context*. She depicted the whole deal. If she knew what the eggs looked like, she showed them. If the caterpillars changed colors with various instars, she showed that. She showed the correct instar munching away on a leaf with the correct damage. She often showed the pupa. If she knew that male butterflies and female butterflies of the same species looked different, she illustrated both.

Exquisite art like this was rare. But *information* like this, remarkably thorough and accurate, was nonexistent. Biologist and historian of science Kay Etheridge's extensive writing about Merian includes a 2010 paper titled "Maria Sibylla Merian and the Metamorphosis of Natural History." The title says it all. Etheridge suggests that Merian was "one of the earliest naturalists to conduct long-term studies on a specific group of organisms."

In her book *Women on the Margins*, historian Natalie Zemon Davis gives readers a tantalizing glimpse of Merian's eloquence by translating part of a passage in the caterpillar book, describing a moth that used a cherry tree. Merian explains that she had seen the moth earlier in her life and been enchanted by its colors. When, by "God's grace," she discovered "the metamorphosis of caterpillars," and finally connected the correct caterpillar to her long-revered moth, she wrote: "I was enveloped in such great joy and so gratified in my wishes that I can hardly describe it."

Hard-pressed to describe her own joy, she described the moth's caterpillar with ecstasy: "They have a beautiful green color, like the young grass of spring, and a lovely straight black stripe the length of the back, and across each segment also a black stripe out of which four little white round beads glisten like pearls. Among them is . . ."

Her rapture continues for several hundred more words.

Merian left her husband and ended up in progressive and prosperous Amsterdam, then as now a center for art, science, and enlightened thinking. There wealthy collectors showed her their butterflies. She saw examples of the fabulous blue morpho butterfly of Central and South America, much loved by butterfly fans even today. She valued the specimens, but also found them frustrating.

Dead butterflies seemed meaningless. What was their context? What plants did they eat? How long did they live? What did they look like flying? What did their caterpillars look like? How long did they pupate? These questions drove her crazy. She needed answers.

So in 1699 she sold her art to raise research funding, boarded a ship, and sailed to Suriname, accompanied only by her twenty-one-year-old daughter. She was *fifty-two* years old. No European—let alone a single woman—had ever done such a thing. Most Europeans heading to the Western Hemisphere were seeking riches. Others were forced to go, while still others went at the behest of king and country.

Merian went just to satisfy her curiosity. The great scientific research expeditions, like that of Charles Darwin, lay well into the future. No one—absolutely *no one*—had crossed the Atlantic by themselves, without sponsorship, in order to do independent field research to answer a scientific question.

Historian Davis has called her "willful," but this kind of risk-taking goes well beyond that. Her fieldwork in Suriname was grueling, even with her daughter to help her capture, nurture, and bring to fruition all the caterpillars she could find. She had expected to stay for five years. After nearly dying from what might have been malaria, she decided to return home after two years.

The 1705 book she produced from this field trip—*Transformation of the Surinamese Insects*—took Europe by storm. The equivalent of a modern Hollywood blockbuster, the book was physically huge, measuring more than fourteen inches by almost twenty-two inches. Merian felt she needed the extreme page size to do justice to the marvels she had seen.

The first copies were meticulously hand-painted by Merian and her daughters. The national library of the Netherlands has one of these and calls it "a major showpiece" of their collection and a "cultural treasure." Sadly, many of these hand-painted copies were disassembled in order to sell individual pages. Other copies, meant for the middle classes, were sold in just black-and-white, at a substantially reduced price.

Perhaps it's not too fanciful to suggest that Merian's attraction to the life stages of Lepidoptera may have been an expression of a secret wish: the desire for change. She was born a woman whose fate was to be a housewife. But she was also a natural-born scientist, with a curiosity that impelled her to find truths rather than accept received wisdom. In the end, she triumphed by becoming who she was meant to be. "Had she painted her own life, it certainly would have mimicked those of her beloved insects," writes insect paleontologist Michael Engel, "a representation of her own metamorphosis, utterly changing the expectations of a woman at the dawn of the Enlightenment."

Her work was, according to Etheridge, "a significant tributary feeding into a growing stream of knowledge, and one whose presence altered the course of the main stream." Many of her images were titillating, illustrating not only butterflies but monstrous insects like the largest tarantula on the planet, and gorgeous plants that yielded luscious fruits not native to Europe, such as pineapples, watermelons, and ripe pomegranates. And frogs and lizards and snakes and birds. And an alligator battling with a deadly snake. Europeans were pleasantly horrified.

"From my youth onwards, I have pursued the study of insects," the book began. "That is why I retreated from the company of people and engaged in this study. In order to practice the art of painting and to be able to draw and paint them from life, I accordingly painted all the insects I could find, first in Frankfurt am Main and then in Nürnberg, meticulously on vellum."

Among her rapturous descriptions was her depiction of the blue morpho butterfly: "I fed this yellow caterpillar in Suriname with pomegranate leaves. On 22 April it attached itself and became a grey pupa, from which this beautiful butterfly emerged on 8 May, being blue and silver and rimmed with brown, interspersed with white half-moons. On the reverse side it is brown with yellow eyes. They fly very fast."

Then she revealed something of her use of the new technology.

"Seen through the magnifying glass, this blue butterfly has the appearance of blue tiles, looking like roof tiles lying in a very orderly and regular manner, being broad feathers like the feathers of the peacock, with a marvelous sheen.

"It is well worth looking into, since it cannot be described."

There is something elemental in the beauty of a butterfly wing. Like the twentieth-century canvases of Mark Rothko or Jackson Pollock, the strong, brilliant colors of a butterfly wing excite our neural pathways in simple, direct, primal ways. We stare. We take a second, a third, a fourth look. What, exactly, are we looking at? We can't quite get a fix on the mercurial colors.

This seems to be what happened to Merian. The blue morpho's "marvelous sheen" bedazzled and frustrated her. Try as she might, she could not reproduce what she saw. Its essence was elusive.

The butterfly's *affect*—its nature, its shifting color, its subjective aspect—could not be pinned down. Instead the insect was like Schrödinger's cat: if you stopped it from flying, the blue morpho's beguiling evanescence vanished. The multitude of colors on the insect's wings were more fleeting than the colors of the rainbow. At one instant the insect looked green. Look again and see purple. Then black. Then the rich blue returned. If you changed the angle of your vision, the iridescence shifted again.

Today we are somewhat accustomed to this constant flow of flash and dazzle because of television and computer screens, which also appeal to our innate neural pathways. Thinking back to Merian's time, her bliss is easily understood. Before screens dominated human existence, such visual experiences were rare.

Yet glutted as we are with an abundance of color in our modern culture, we, too, are still hypnotized by seeing a blue morpho. Go to any butterfly house. The blue morpho will be the most popular insect there. Toddlers chase after them with irrepressible ardor. They *want* that butterfly.

Bush pilots flying over blue morpho territory can pinpoint this butterfly from hundreds of feet above the jungle because the flamboyant flashes of the male insect's raffish blue are so intense. (Females are also blue, but less blatantly, glaringly so.) Yale ornithologist and butterfly lover Richard Prum told me about a walk he had one fog-filled March morning on the east slope of the Peruvian Andes, just beneath the Inca city of Cusco.

"It was a magical altitude for morphos," he said. All of a sudden, as the day warmed, the fog lifted and he was flash-mobbed. Dozens and dozens of the butterflies suddenly began flying from ten to twenty feet above his head. Their flashing, dazzling brilliance popped and popped over his head.

If Merian had had a scanning electron microscope, she would have learned that the morpho's opalescent colors, which so intensely shock our brain's visual pathways, derive not from pigment, as she assumed, but from the very *structure* of the scale itself. She was a victim of physics: the insect's transcendent, shimmering aura could never have been replicated with her watercolors.

Blue is a weird color. Common as it is—blue sky, blue ocean—blue *pigments* are unusual. In Merian's time, artists choosing to work with blue colors literally paid a heavy price, because the color was hard to find and extremely costly. Usually it came from lapis lazuli, a semiprecious stone.

Most blues in nature come not from pigments, but from the surface structure of the object we're looking at. This is true even when we look at blue eyes: there is no blue pigment in the blue eye. Instead, there are structures that scatter most light except for the blue wavelength.

The most common everyday experience we have of *structural* color is seeing colors shift and change in soap bubbles. Dip your circular wand into a jar of soap, blow into the air, and watch the colors in the bubbles shift and change as the bubbles float away: you are enjoying the phenomenon of structural color. Colors deriving from the structural scattering of light profoundly impact our visual neurons.

A rich blue sky—the result of crystalline structures in the atmosphere—gets our attention and makes us feel like a million dollars. One long-ago summer day in southeastern Vermont, the dreary Green Mountain skies, filled with clouds and drizzle, got the best of me. I spotted a far-distant patch of blue sky, jumped in my car, and chased it across the state, all the way to the New York border. I never did catch it.

The first time I saw a blue morpho, I was similarly entranced. I felt an immediate uplift of pure excitement followed by a joyous, greedy delight. I wanted *more*. The blue wings held me transfixed, as though the butterfly had cast a spell. The blue seemed more vivid, ostentatious, almost alive. I couldn't get a fix on the precise shade. The soap-bubble-like blue kept dancing. Was it more green? Was it darker? Was there actually some black in there? Its hue vibrated.

This, it turns out, is the response that nature intends. The more you are flashed and dazzled, the more time the butterfly has to escape.

Michelangelo attained a similar effect in his *Holy Family*, which hangs in Florence's Uffizi. The blues of Mary's dress vibrate and shimmer. Seeing this work, I was trapped, just as when I saw a blue morpho or when I chased that patch of blue clear across Vermont. It was as though a hypnotist were waving a watch back and forth in front of my eyes.

The key to the blue morpho's razzle-dazzle is the shape of the butterfly's scales.

Scales—a lepidopteran innovation—cover not only the wings, but also the insect's body and legs. As the insect develops in the chrysalis (butterfly) or cocoon (moth), a single living cell divides into two separate living cells. These cells ultimately become (1) the socket holding the scale and (2) the scale itself.

On a flying butterfly, the scale that we see is dead. But in the chrysalis this same scale was once a living cell, with all the normal parts of a cell—nucleus, cytoplasm, and so on—all encapsulated by a multilay-ered and flexible cell membrane. Think plastic bag holding fluid with lots of stuff floating in the fluid. Our own bodies are composed of cells like these.

As the butterfly develops, the living cells die. The interior com-ponents disappear. The cell membrane, however, remains. This once-malleable plastic bag–like surface hardens, but not before it forms into structures that reflect light in unusual ways.

In most Lepidoptera species, the dead scales are hollow inside. They lie on the wing in orderly fashion, parallel and in rows, like shingles on a roof. Merian mentions the orderliness of the morpho's wing scales, so we know that her technology allowed her to see at least that much.

Made of chitin, a hard substance composed of long sugar chains, wing scales are minute in size: the human eye registers them as dust or powder. Butterfly scales are so small that people who work a lot with Lepidoptera wear face masks in order to keep from inhaling the scales and developing lung problems.

The scales are not tightly fastened onto the wings. When enough scales dislodge from a butterfly wing, the wing may look transparent. Scientists suspect that the scales may help lift the butterfly when flying. Nevertheless, butterflies that have lost many scales can still fly quite well. When people say that a butterfly looks old and worn, they are refer-ring to the diminished, dull look that comes from a heavy loss of scales.

The shape of the scales and the patterning of the layering vary from species to species. Some species have scales that are long and hairlike, while others have scales that look like the blades of old-fashioned canoe paddles.

Wing scales of butterflies are multipurpose. Because they detach so easily from the body, they have defense value: if an insect gets caught in a sticky spider web, the animal can escape easily by "slipping" its scales, the way you might leave your jacket behind were it snagged on barbed wire.

The color of butterfly scales also helps either to draw attention to the butterfly or to hide the butterfly. When the blue morpho spreads its wings, the shimmering blue is there for all to see. As they fly flamboyantly in the sunlight, you can't help but notice them.

But the butterflies are in fact *hiding* this way—hiding in plain sight. They are using shock and awe: the unstable light reflected off the scales is confusing. Our eyes can't quite get a fix on exactly what we are looking at. We are not the only predators who take second, third, and fourth looks. A predator like a bird may be startled and have trouble focusing. That lost instant of focus may be just long enough for the butterfly to escape.

The blue morpho has other color-related lines of defense as well. When the wings are folded up, perhaps when lying on dead leaf litter, the insect blends in thoroughly with the natural background. On the undersides of the wings are scales of dull browns and tans and blacks. Eyespots—much like the eyespots of male peacocks—are prevalent. There may be as many as five eyespots, using the "safety in numbers" principle to warn away birds. On these undersides, there is nary a glimmer of blue. If the insect is resting on the bark of a tree, it's almost invisible. You would never guess the marvelous color hidden on the other side of the folded wings.

Many butterflies have this "dual-personality" option. The Indian dead leaf butterfly, *Kallima inachus*, looks astonishingly like a dried-up old leaf—when its wings are folded up and the butterfly is resting. When the insect opens its wings, it's another story entirely. Blue colors flash and dazzle in the sunlight, along with wide stripes of gaudy orange.

The basic artistic philosophy of butterflies seems to be: if hiding doesn't work, try in-your-face ostentatiousness. Two centuries after Maria Sibylla Merian, this two-faced strategy would find its way into the center of the debate over evolution, with Darwinists claiming it as proof of evolution—and with anti-Darwinists claiming that the intricate beauty of butterflies had to have been designed by a deity.

What Maria Sibylla Merian could never have seen with her technology were the minute details of the surfaces of butterfly scales. This is something science has only recently grasped. Among a select group of wing scale experts the discovery has made big news. Indeed, they have teamed up with engineers to study these details in order to make quantum advances in computer speed and energy efficiency.

Nipam Patel grew up in Texas. He began collecting butterflies at the age of eight, and by the time I visited him, thirty years into his professional career, he had amassed more than 50,000 specimens, putting him right up there with the Big League collectors of the Victorian era. In 2018, Patel was asked to leave behind his laboratory at the University of California, Berkeley, to head up the prestigious Marine Biological Laboratory in Woods Hole, Massachusetts, near where I live. As a condition of acceptance, he required the 140-year-old institution to create something new: modern housing for his butterfly collection. Without that, he explained, he would have to decline the invitation.

Patel is an expert in embryology: how living things develop from eggs to adults. As such, he is a scientific descendant of Maria Sibylla Merian. Patel's lab has devoted years to understanding the development of the wing of the blue morpho. Researchers there have discovered a way to spy on the insect's wing as it develops to maturity. Time-lapse videos show the insect's "proto-wing" transforming into the brilliant wing we see when the insect is flying.

One of Patel's current passions is thinking about the physics of beauty, he told me when I went to visit him in Woods Hole.

"Funny tricks happen with light," he told me as we talked.

He mentioned the soap bubble effect and the colors of the rainbow. I added that I'd seen something similar in a sheen of oil.

Then he showed me pictures of Christmas trees—or at least what this group of butterfly-mad scientists are currently calling "Christmas trees." In Patel's lab and in other labs around the world, researchers have managed to slice morpho scales crosswise, and, using electron microscopes, have found that the scales have a particular, orderly shape that reminds them of the outline of a pine tree.

It is this pine-tree structure—extant on the nanoscale, precise and orderly and frightfully fine-tuned—which produces the color. To grasp this bold, brilliant, bizarre fact, remember that the scale starts as a living, pliable material containing the guts of a living cell.

Remember that the cell membrane starts out like a pliable plastic bag. The butterflies manage to cause this "plastic bag"—the cell membrane—to bend into these specialized shapes that reflect light in these highly specific ways. Using proteins inside the scale cell, physical forces cause the cell membrane to contort and bend predictably.

I talked to Yale's Richard Prum about this.

"In the morpho," he said, "the 'garbage bags' begin to form spiky long ridges that start to get folds in their surfaces."

I tried to think of an analogy to this strange phenomenon, but I failed miserably. What happens when the scale of a butterfly turns from a shapeless and malleable living membrane to a dead and rigid structure with specific nanoscale outlines is like nothing else I know of.

As the membranes die, the scales of most butterfly species form ridges. Think of them as the kinds of ridges you see on corrugated steel roofs—regular, repeated, orderly. It is these repetitions that help to manipulate the light. Then the corrugations themselves bend and stretch into the pine-tree shapes.

On a blue morpho scale, light from the sun is bounced around and various wavelengths are "thrown away," or scattered. Only one wavelength—blue—remains organized enough to be effectively reflected back to the observer.

This is one reason why we find this color so exciting: This blue is *pure. Unadulterated. Clean. Fresh.* Colors created by pigments do not have this quality. They can be almost dull in comparison.

You might be thinking: This is nice to know, but what's the point? Why spend so much time and money discovering minutiae like this? There are practical as well as esthetic reasons to study such things. It turns out that discovering the structure of blue morpho scales may well help save many a human life.

Radislav Potyrailo, a physicist/chemist/biologist/electronics wiz originally from Ukraine, designs vapor-selective sensors that can detect poisonous gases in the atmosphere. His work has all kinds of practical applications, such as helping people with asthma, or detecting potentially poisonous gases emanating from a volcano, or detecting poison released in an underground subway.

Potyrailo had access to a variety of such sensors, but he didn't like them much. Either they were inexpensive and didn't work well, or they were too expensive and heavy and awkward to carry around.

"You can't put a shoebox or a laptop in your pocket," is the way he explained it to me.

This is true. I had to agree with him on that.

There are smaller sensors already on the market, he continued, but they don't work well. An asthma sufferer with one of these may end up being alerted to danger, only to find that the sensor was detecting aromas emanating from something innocent, like cheese.

He wanted to create something that would combine the efficiency of the large sensor with the convenience of the small one. A lecture by one of his colleagues about the shape of blue morpho scales gave him an insight. Not a special fan of butterflies, he hadn't thought much about their scales. But in the lecture, a lightbulb went on. His colleague showed the scale's pine tree–like cross section.

He was "bio-inspired," to borrow from his lexicon. He took the "design rules" of the scale (thank you, evolution) and applied them to his own designs. "I was able to compare our new butterfly-inspired structure, and to our surprise it worked better," he explained. "We mimicked the scale design. After that, we went beyond our nature-inspiration. It unleashed our thoughts in different directions."

Other butterfly species use the "trash bag" technique to create different shapes and structural colors. Currently scientists are particularly interested in scales that shape themselves in a way that projects a vividly metallic green. The discovery of this particular shape in a butterfly wing scale sent shock waves around the world—at least in scientific domains.

The structure had been *theoretically* imagined decades earlier, in 1970. After years of intense mathematical contemplation, NASA physicist Alan Schoen, seeking to invent a new lightweight material, invented a groundbreaking idea: the *gyroid*. It was a pretty cool concept. Gyroids,

as Schoen dreamed them up, were strange mathematical surfaces—
crystalline three-dimensional structures that allowed for the almost infinite flow of energy.

To picture a gyroid in your mind's eye, imagine a honeycomb, but
in three dimensions. Then imagine that you could slide from one "unit"
of the three-dimensional honeycomb to any other unit just by slithering
through the infinitely connected labyrinths.

Gyroids, Schoen theorized, were simply sophisticated geometrical
objects that could expand and grow as necessary. They could do so with
great economy, as they use the minimal amount of surface material to
get the job done. Gyroids, as Schoen imagined them, could exist on a variety of scales, from quite large to infinitesimally tiny. Schoen was thinking about them from the point of view of space travel, which required
tough but lightweight materials.

His idea took hold. San Francisco's science museum, the Explorato-
rium, built a human-size gyroid for kids to climb through. Technology
companies researched Schoen's idea, hoping to create better solar cells
and communications systems. The idea seemed to emanate purely from
human inventiveness and was considered revolutionary.

Instead, it turned out to be *evolutionary.*

Butterflies had come up with the gyroid tens of millions of years
earlier. Green hairstreaks shape the surfaces of their scales into the gyroid configuration, manipulating the flow of light by isolating a certain
wavelength of light. All other wavelengths have disappeared, scattered
to the four winds, so to speak. All that is left for the eye to receive is this
one unique energy wave, which our eyes perceive as a fabulous metallic
green.

The gyroid is in essence an optical filter. Think of Newton's prism
from Maria Sibylla Merian's time, but instead of separating light into
a rainbow of colors, the gyroid neutralizes all but that one special hue.

The hairstreak's gyroid has been called by one team of scientists
"one of nature's most symmetric, most complex and most ordered structures." A team of Australian scientists has already mimicked this butterfly's gyroid to create a human-made 3-D structure that they hope will

eventually be developed for use in computer technology by replacing soldering boards with channels of manipulated light energy. Others are applying what they've learned to improve anti-counterfeiting logos.

All this sounds so complex that you might imagine a particular species would require eons to evolve different colors. But it turns out that these color changes can happen almost instantaneously. A team of Yale researchers found in 2014 that butterflies can change color in the blink of an eye, evolutionarily speaking. The team took a species of butterfly with mostly dull brown wings and bred it to a related butterfly that had some purple on the wings, but not much. After only one year—six generations of purple mating with purple—the dull browns had given way to exciting purples.

Sometimes scale color changes with the seasons. A small brown African butterfly that lives on the savannah has brightly colored eyespots during one time of the year, but its offspring have dull colors that help them survive during the six-month dry season. Marvelous though this is, animal use of structural colors may date back hundreds of millions of years and be rather run-of-the-mill in nature. Some scientists propose that dinosaurs may have used structural colors, along with pigments, to color their fabulous feathers.

So what does all this have to do with Charles Darwin and the scientific revolutionaries who agreed with his ideas?

Five

HOW BUTTERFLIES SAVED
CHARLES DARWIN'S BACON

It was snowing butterflies.

Charles Darwin, The Voyage of the Beagle

Charles Darwin knew about Merian. Although he did not mention her in his copious letters, he owned an encyclopedia that reproduced at least one of her artworks. By the time Darwin was born, the knowledge she had spread throughout Europe had become widely accepted. Several of Darwin's colleagues revered her.

Darwin as a young man circled the globe for five years on a collecting and exploring ship financed by the British government. His voyage on the HMS *Beagle* brought him to various locales along South America's coastlines in 1832 and 1833. His father made sure young Darwin had plenty of money for whatever luxuries might be available to him as he explored. Unlike Merian, he was well looked after.

For most of his life, Darwin seems not to have been overly interested in butterflies. As a student, when on outings while the other boys collected butterflies, he searched for beetles. On his exploration voyage, he seemed nonplussed by the Lepidoptera, in contrast to other Europeans. On a walk through the forest near Rio de Janeiro in 1832 Darwin saw "large and brilliant butterflies, which lazily fluttered about." This was hardly the ecstasy of a butterfly fanatic, or the ecstasy he himself had experienced as a younger man when he discovered an unusual beetle under a rock.

His lack of interest was more than made up for by three subsequent European explorers, all Victorian butterfly fanatics whose research into butterfly wing patterns would provide the first contemporary real-world proof that Darwin's theory of evolution was a factual description of how nature works on our living planet. Evolution, these scientists would show, is a ceaseless process, one that occurred in the long-distant past, but that also operates in the present and will continue to operate on into the future.

Breathless from reading Darwin's 1839 round-the-world adventure story *The Voyage of the Beagle*, as well as adventure travel books by Alexander von Humboldt and by American butterfly addict William Henry Edwards, two young acquaintances decided to sail from England to South America in 1848 to try to earn a living as collectors of natural history specimens.

The duo—Alfred Russel Wallace, age twenty-five, and Henry Walter Bates, age twenty-three—came from families that had once been middle-class but had suffered financial setbacks. Neither had anything but a rudimentary education, yet both were destined to rank among science's most famous names—in great part because of their fascination with butterflies. The men met in a public library in Leicester in Britain and immediately became fast friends, reading the same books and discussing mutual scientific interests. Neither faced a bright future. Bates was apprenticed to a hosier. Moreover, 1848 was a year of revolution across Europe. Adventure overseas looked like a better bet to both.

Wallace and Bates spent their first year in South America collecting together, then went their separate ways. Wallace returned to England in 1852. Tragically, most of his specimens were lost at sea when his ship caught fire. Rescued, he vowed never to go abroad again when he got home to England. Soon thereafter he was off to the Malay Archipelago, where he encountered a butterfly whose beauty so emotionally overwhelmed him that he spent the rest of the day with a stress headache.

Laid up in the East with what might have been malaria, he conceived of an idea that would be explained in a short paper entitled "On the Tendency of Varieties to Depart Indefinitely from the Original Type"—essentially his own independently derived theory of evolution. The date of this paper, written on an isolated Indonesian island, is 1858—a full year before Darwin published *On the Origin of Species*.

Unbeknown to each other, the two had been working on the same problem: Are species static and immutable (unchangeable) or do they evolve (change) over time? The *scala naturae* required immutability: everything had its place. Evolution allowed for ongoing change and flexibility rather than rigidity: if everything depended on everything else, then there could be no innate "superiority."

Bates remained in South America for eleven years, returning to Britain in 1859—160 years after Merian first set sail. Eighteen-fifty-nine was a remarkable year, one that set the stage for changes that would rock the foundations of Western culture, as the scientific revolution begun in Merian's era culminated in breakthroughs that would create the world we live in today. Obscure New Englander Moses Farmer turned on the world's first electric lightbulb for his wife on the mantelpiece of the family's small home north of Boston, marking the beginning of our electrified planet. John Brown raided the Harpers Ferry Armory, setting off the war that would finally end American slavery. Charles Dickens's *A Tale of Two Cities* warned of dangerous times to come, unless the world's wealthy addressed the needs of the poor.

And Darwin published *On the Origin of Species*, which ultimately ended the rule of *scala naturae*. Thus *Origin* was among the most politically volatile books ever published. If species could evolve and adapt in accordance with the planetary times in which they lived, then where was natural hierarchy? And if there was no natural hierarchy, how was society to be organized? It's not hard to understand how evolutionists were seen by some as being in league with the devil. In the year 1859, Maria Sibylla Merian's science finally reached its true potential. If life was not a hierarchy but a net, who would rule whom? If rulership did not derive from the divine, then who would set standards of behavior?

Darwin seemed to some to be a Pied Piper leading civilization to the precipice. Hostility materialized immediately.

In a review of *Origin*, the devoutly religious entomologist Thomas Vernon Wollaston claimed that all species were immutable and created by God. The existence of *butterflies*, he wrote, proved Darwin wrong: "We cannot conceive that such marvellous perfection of painting as, for instance, the tints of certain butterflies (which are blended together with such nicety and consummate skill, in accordance with the laws of colouring, as to surpass an artist's touch) could have been brought about through mere correlation with a change in some other part of the organism. . . ."

This would not turn out well for Wollaston.

Far from being immutable, butterflies were about to become a prime example of evolutionary theory. Darwin was deeply disturbed by Wollaston's hostility. They had discussed the issues collegially several times, and he took his colleague's butterfly-based attacks personally. Darwin was a quiet nonbeliever, more than willing to respect others' religious beliefs. Although *Origins* was far more revolutionary than *Das Kapital*, Darwin, unlike Marx, was not by nature a revolutionary.

He just enjoyed thinking things through to their logical conclusion. Even at the end of his life, when people applauded if he merely entered a room, he much preferred the study of worms—the subject of his final published book—to basking in the limelight. When I walked around Darwin's estate, Down House in the village of Downe, I saw that these worm experiments still litter his grounds. He was a scientific explorer, a putterer, and an author to the very end. His final book: *The Formation of Vegetable Mould, Through the Action of Worms: With Observations on Their Habits.*

Another flashy title.

As the controversy raged, younger scientists like the volatile Thomas Henry Huxley, a pugnacious man who described himself as Darwin's "bulldog," spoke up for Darwin, but Darwin himself took refuge in a health spa, seeking treatment for his ever-present health issues. (He had never gotten over the early death of a beloved daughter.) He hoped to let the battle play out.

This did not happen. By the time he finished his water cure, the controversy had strengthened to gale force.

Then Henry Walter Bates and his butterflies came to the rescue.

In March 1861, Bates wrote to Darwin that he had found evidence that some butterfly species change their wing colors to mimic the wing colors of other butterflies. This, Bates suggested, helped them avoid being eaten. "I have an immense number of facts on this subject," Bates wrote in closing. "Some of these resemblances are perfectly staggering, —to me they are a source of constant wonder & thrilling delight."

Facts! Darwin's eyes likely lit up. Just what he needed. Had he been part of the Victorian butterfly-collecting crowd, he might have seen the truth about butterflies. Since he hadn't been interested, Bates had to point out the facts for him. Bates had found in the Western Hemisphere a group of butterflies that, in Darwin's words, wore "deceptive dress." Bates himself called this group "counterfeits."

Essentially, these butterflies were scam artists. They pretended to be what they were not. During his eleven years in South America, Bates had kept in-depth records of everything he saw, including the Lepidoptera. He noticed that he often saw large numbers of particular types of butterflies flying together. He also noticed that there seemed to be another species that looked similar flying along with the larger group. The second, less common species was, oddly, colored quite like the majority species.

The majority species turned out to be unpalatable. A predator who took a bite would either regurgitate the meal or actually die. The minority species observed by Bates, however, was highly palatable. Yet predators avoided the minority species just as though it were, like the majority species, inedible. In other words, the minority species was *faking it*. It was surviving by blending in with the badass majority species.

Coincidence?

Bates thought not.

Elsewhere Bates found other highly palatable butterfly species surviving by looking like other unpalatable species. It would eventually turn out that some butterfly species could change their coloration quite quickly, in only a few generations, depending on their other butterfly colleagues. As Maria Sybilla Merian had shown two hundred years earlier, context was the key.

"I think I have got a glimpse into the laboratory where Nature manufactures her new species," Bates wrote.

Elated, Darwin agreed.

In the face of Wollaston's declaration, this truth must have been delicious to Darwin. He encouraged Bates to publish. When the paper came out with a rather bland title—"Contributions to an Insect Fauna of the Amazon Valley"—Darwin feared this essential piece of supporting evidence for his theory might be missed by scientific leaders.

To overcome the nondescript title, he surprised his colleagues by stepping forward to author comments on the paper. This was not something he usually did, but he worried that the paper might have been "overlooked in the ever-flowing rush of scientific literature."

With Darwin in the saddle, Bates's paper would not be overlooked: "The main subject discussed is the extraordinary mimetic resemblance which certain butterflies present to other butterflies belonging to distinct groups. To appreciate the dissimulation practised by these insects," he continued, just look at the "beautiful plates" published in the paper. "Travel a hundred miles" and you will find other examples of "mockers and mocked," he gushed. (The mockers were the con artist butterflies who mimicked the genuinely toxic butterflies.)

"The mockers and mocked always inhabit the same region; we never find an imitator living remote from the form which it counterfeits."

Darwin continued: "Why then, we are naturally eager to know, has one butterfly or moth so often assumed the dress of another quite distinct form; why to the perplexity of naturalists has Nature condescended to the tricks of the stage?"

Darwin, of course, knew the answer: "due to the laws of variation!"

I.e. . . . *evolution*.

By changing its scale colors, the minority insect had a better chance of surviving.

You can just hear Darwin licking his chops.

Take that, Wollaston!

By all accounts, Charles Darwin was a kind rather than vindictive man, but he could not resist crowing over this triumph. Perhaps, given the vehemence with which he had been attacked, he can be forgiven this peccadillo.

Darwin was also pleased to find that, unexpectedly, evolution could be seen in the present. He had not expected that. Initially he had written: "We see nothing of these slow changes in progress, until the hand of time has marked the long lapse of ages." He believed that evolution "always" acted quite slowly. But after hearing from Bates and Wallace and so many others, he changed "always" to "generally."

I personally suspect that he was pleased to be able to do so.

Once citizen scientists began looking, they found plenty of examples of this kind of mimicry. Batesian mimicry—in which a harmless species takes on the coloring of a dangerous species—turned out to be an everyday occurrence: "Butterflies were destined to become evolution's most elegant practical support," wrote the Darwin biographer Janet Browne.

A third naturalist-wanderer, the German Johann Friedrich "Fritz" Müller, discovered in the forests of South America yet another type of most marvelous mimicry. Müller found that two unpalatable butterfly species might, over time, *mutually* change their wing scale colors and patterns to more resemble each other. In other words, they compromised. They formed a kind of Mutual Protection Society, based on safety in numbers. If a predator tasted only one unpalatable butterfly, then, Müller showed, other butterflies of similar appearance, regardless of species, were less likely to be victims of that predator.

Again Darwin was delighted. He had Müller's book in German on the subject first translated for him personally. Then he financed an English edition for the public at large; an "insectivorous protection racket," adding that the discovery thrilled Victorian society: "Much of the animal world, it seemed, was driven by forgery."

Today we understand that color—scale color, fur color, hair color—is often a matter of simple genetics. A gene turns on. A gene turns off.

Sometimes it's only a matter of temperature. Other times, it's a matter of blending in with the crowd. Or of standing out from the crowd.

But remember: in Darwin's day, no one knew about "genes" and the sometimes simple process of biological change. One of my personal favorite examples of how quickly evolution can bring change comes from the avian world. Charles R. Brown and Mary Bomberger Brown collected thirty years of data on cliff swallows killed by speeding cars in southwestern Nebraska. They found that population numbers dropped over those three decades. Then they found that surviving birds began increasing in numbers. But the surviving birds had changed. Their wings were a few millimeters—roughly a tenth of an inch—shorter. The slight change in wing length allowed the birds to get out of the way of oncoming cars more quickly.

The best-known modern example of Lepidoptera microevolution is the wing color changes of the British peppered moth. In the early 1800s around preindustrial Manchester this moth was light-colored with darker spots (hence the "peppered"). This camouflaged the insect against the light-colored lichens and tree bark on which it rested. When industrialization exploded and the region's air turned toxic with coal pollution, the light-colored version of the moth disappeared. The moth now was almost completely dark—mimicking the soot-covered tree bark that had become common. When anti-pollution laws were passed, the air cleared and the light-colored version became common again.

Only a few years ago, geneticists discovered that this quick-change artistry was due to one particular mutation in one gene. What Darwin and others believed to be enormously complex and seemingly almost miraculous—the changing of colors on a butterfly wing—we now know to be rather simple.

Indeed, changing colors may be the norm. Recently in a lab at Yale University, a team of scientists bred boringly colored butterflies with the appropriate name of squinting bush browns. By persistently pairing individuals of these species who seemed to have a bit of blue or purple, over the course of six generations the scientists came up with brown butterflies that had purple streaks on their wings. "It seems to be incredibly easy to evolve these new colors in butterflies," study scientist Antonia Monteiro told National Public Radio.

Even caterpillars are experts at camouflage and mimicry. Plenty of caterpillars imitate the dead leaves of the plants on which they feed. Other caterpillars mimic bird droppings, twigs, rocks, tree bark . . . the list is endless.

Yale caterpillar expert Larry Gall once showed me a page of caterpillars photographed on their host plants. In a *Where's Waldo* exercise, he asked me to find them. I couldn't find a single one.

Other caterpillars, though, have entirely different strategies. Like butterflies, they adopt flash and dazzle.

Our eight-year-old granddaughter found one such example in our front yard. In mid-August, prowling through our butterfly garden (by now, the whole of our front yard), she found a large bright-green caterpillar with yellow snake eyes. This was the final instar of the spicebush swallowtail caterpillar. Now large enough to be easily seen, the mimicry of the glaring eyes of a snake might be enough to deter most predators.

When I saw those eyes, I immediately jerked back.

It didn't work with Elena, though.

She loves snakes.

When Bates and Wallace disembarked from the ship that had carried them across the Atlantic from dreary Liverpool to sunny South America, they found themselves in an entirely different world. Even as Bates suffered in South America—he was often destitute and several times was seriously ill—the place thrilled him. He did not miss the cold weather. From his earliest days on the continent, he reveled in seeing seemingly infinite numbers of butterflies. In Britain there were few species, but in South America he could sometimes see hundreds of species in only one day's ramble.

His energy and self-discipline were astonishing. The human brain tends to give in to heat by slowing down, taking it easy, and becoming more relaxed. I know mine does. But Bates just kept going, week after week, month after month, for eleven years. From time to time, he kept company with others who shared his European culture, but most often he was either on his own or with the many indigenous people with whom he made friends.

By the time he returned to Britain, he had collected almost 15,000 animal species, including 8,000 that were new to science. Many of these he had sent back home long before he himself boarded a homeward-bound ship. One butterfly had been named after him: *Callithea batesii.* He was prey to the same temptations as Maria Sibylla Merian and often endangered himself in order to collect. "I have collected every day a splendid box of butterflies besides other things," he wrote to his brother, "always taking something new; and in spite of the furious heat of the sun and great fatigue, enjoyed myself amazingly."

Like Merian, he became deathly ill and only barely survived.

And like Merian, he was intoxicated by the blue morpho's beauty.

"It is a grand sight to see these colossal butterflies by twos and threes floating at a great height in the still air of a tropical morning. They flap their wings only at long intervals, for I have noticed them to sail a very considerable distance without a stroke," he wrote in his memoir.

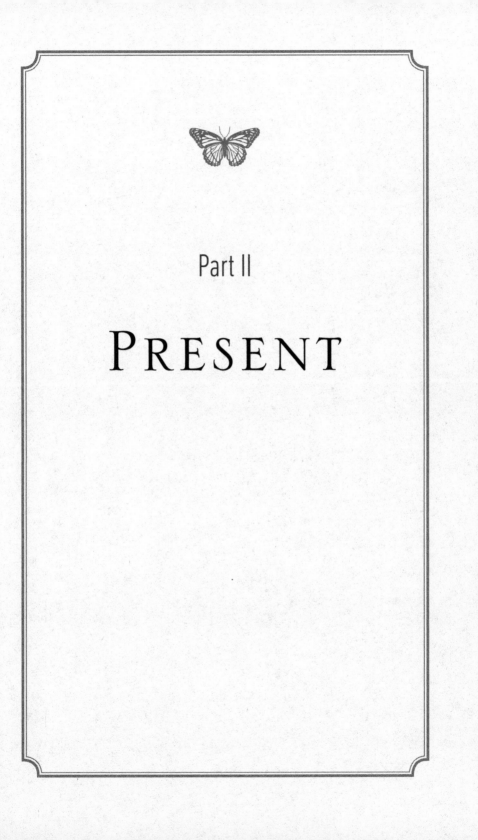

Part II

PRESENT

Six

AMELIA'S BUTTERFLY

. . . these are flowers that fly and all but sing . . .

Robert Frost

In the fall of 2016, five-year-old Amelia Jebousek brushed her long hair out of her eyes for about the umpteenth time. She could barely stand still with anticipation. Finally, it was time. Raising her hand high, she released her prize.

Against the cerulean sky of Oregon's fertile Willamette Valley, the child's butterfly hesitated. Then she spread her wings. Flying up to a nearby branch, she took her bearings. For the first time in her short life, the creature surveyed the universe in which she was destined to fly, and where she would ultimately come to symbolize all that is best in science and in the human heart.

"Look about you at the little things that run the earth," the great biologist E. O. Wilson wrote in *Half-Earth*. When I read this, I assumed his elegant phrase was mere poetic license. Obviously, it was we mammals who ran the earth. Anybody could see that.

But now, as I write this, after two years of chasing butterflies, I have come to understand what he meant. Amelia's monarch was gifted, behaviorally flexible, intelligent—more than worthy of the special scientific research program in which she was enrolled. Orange and black, weighing only two-hundredths of an ounce (less than a paper clip), she belonged to a group of winged beings whose beauty has mesmerized humanity for thousands of years.

Her antennae, essential for migration, were only as thick as a few strands of human hair. Her wing scales, the functional equivalent of a bird's feathers, were so infinitesimal that they looked like nothing more than a mote of dust when separated from the wing. Yet they were so intricately shaped that only electron beam microscopes could reveal their staggeringly complex structure.

Amelia's butterfly seemed so frail as to be feeble. She wasn't. This particular insect would perform remarkable feats. Her eye-opening achievements would profoundly improve our understanding of monarch behavior, and indeed of the insect world in general. Our human would be greatly advanced by what we learned from her.

But all that lay well into the future. For now, she was busy adjusting to life outside the chrysalis. Finding a nearby resting place, she spread her wings, creating a natural solar panel that would warm her body.

Then she flew higher, rising on the warming winds until she was truly airborne. As her wings oscillated, her intricate brain, with its complex internal navigation compass, mysteriously retained experiential memories from her life as a caterpillar. Accordingly, she would look and behave differently than her immediate parents, summer monarchs who flitted away their time lofting from flower to flower without any clear direction.

Amelia's butterfly would be a bit richer in color, a bit larger, a bit more suited to long-distance flight. Unlike her immediate parents, she was physically equipped to migrate all the way to the California coastline. She started off on her mission almost immediately.

When she arrived at her destination, as with all migrating monarchs, she would go through a process scientists call "overwintering." She would cluster together with thousands of other monarchs on tree branches, trying to stay warm in the cool weather.

During that period, which lasts through the winter until late February, she would not eat much, and would instead rely on fat reserves she had stored in her body. When spring came, she would leave her roost and head out to find nectar and milkweed plants, where she could lay eggs. The butterflies from those eggs would fly a short way north, and then lay more eggs. And on and on, for three or four or five generations, until the fall came again and her descendants undertook this same migratory journey.

This was her unconscious goal as she made her way southwest. Fine-tuned to see subtleties of light and color invisible to us, her strange compound eyes took in the wide, fertile valley below. She was marvelously evolved to make use of the wetland meadows where her parents had meandered in the warm summer air, enjoying protected expanses rich with wildflowers and patches of milkweed. They had mated, laid eggs, and died. As butterflies, they would have lived a month at most.

This butterfly's life would be different. The angle of the sun, declining day lengths, and her own biology had prepared her to live longer. Months longer. A member of the "Methuselah generation," she had an extraordinary responsibility: the survival of the species.

Choosing a path that monarchs have likely taken for millennia, she climbed high into the sky to ride the southward winds. But her trip would not be precisely the same as that followed by her distant ancestors. The world into which she had emerged had changed enormously. Then, her valley was inhabited by the first peoples of North America—hunter-gatherers who showed up 15,000 years ago or even earlier. Like all humans, they made their mark on the land, but did so in limited fashion.

Now a substantial highway cut through the length of the valley. The expanses of nectar-rich wildflowers and wetlands had been replaced by monocultures of farm fields, vineyards, plantations of Christmas trees, extensive orchards of nothing but hazelnut trees. The areas her predecessors had used no longer existed.

Fortunately, Amelia's butterfly would be genetically equipped with plenty of behavioral flexibility. Even though the changes in the valley over the previous century were extreme, she would still be able to follow the ancient cues embedded in her brain to arrive at her destination. This is the splendor of evolution, what Darwin called its "grandeur." Amelia's butterfly would excel in survival skills.

But the child's butterfly would also triumph in another way, one that would give this insignificant-seeming little creature a thoroughly modern role. Amelia had placed a nearly weightless polypropylene tag on the monarch's wing. Smaller than a thumbnail, the tag notified anyone who saw this insect: Kindly relay your observation of this insect's location, dear sir or madam, to the biologist in charge of this monarch monitoring project. An email address was provided.

The tag worked. Over the next several months, various individuals took digital photographs of Amelia's butterfly and emailed them to project scientist David James. The entomologist then shared her story.

On the simplest level, the life cycle of the monarch butterfly is the life cycle of all butterflies and moths. As Maria Sibylla Merian showed hundreds of years ago, the female lays eggs. Those eggs mature into caterpillars. The caterpillar emerges from the egg and begins eating. Over time, the caterpillar grows ever larger, shedding its skin several times in the process. These stages are called "instars." Then the caterpillar becomes a *pupa*—naked, and also called a chrysalis, in the case of butterflies; wrapped in a cocoon in the case of moths. When the time is right, the insect emerges from its hideaway as a full-fledged flying being.

But these are only general rules. Of the roughly 20,000 or so species of this insect group, each is finely adapted to the ecosystem in which it lives, and each therefore has a unique life cycle. Indeed, even within species, individual butterflies may live different lives.

The monarch is an excellent example of this. North America harbors two major populations of monarchs: monarchs that live east of the Rocky Mountains, and monarchs that live west of the Rocky Mountains. In general, the western population migrates to the Southern California coast for the winter. The eastern population migrates south, sometimes all the way to Mexico. But within specific populations, some monarchs migrate while others do not. Most migrating females cannot reproduce. It used to be said that they "never" reproduce, but in only the last few years, that myth, too, has been upended. If environmental conditions are welcoming, migrating females may well lay eggs.

Monarchs are a "weedy" species. This is not an insult. It's a compliment. It indicates hardiness. Some butterfly species—for example, the small blue butterflies we will meet later in this book—are so highly adapted that when their system is disrupted the least little bit, their future as a species is imperiled.

Not so when it comes to monarchs. Monarchs are survivors. There are populations of monarchs that live year-round without migrating, in some southern states such as Florida. There are monarchs in Cuba, in Mexico, in Spain, in Guam, and even in Australia. Some Australian monarchs migrate as the seasons change, while others—perhaps even others in the same population group—do not. Scientists do not know for certain why this is so, but as you'll see later in this book, they come closer to answering this question with each passing year.

The monarch does, however, have one nonnegotiable requirement for survival: *the milkweed plant*. This is an absolute. No milkweed, no monarch. Fortunately for the monarch, there are roughly 200 species of milkweeds worldwide. These hardy plants, which we used to think of as weeds that needed to be destroyed, bloom with astonishingly beautiful flowers. Their hues range from simple whites to glowing oranges, reds, yellows, pinks. . . . But it's not milkweeds' flowers that the monarchs require.

It's the poison in their leaves.

I had known the popular monarch story for decades, ever since the saga landed on the August 1976 cover of *National Geographic* magazine. Each fall, tens of millions of orange butterflies fly from points north on the North American continent south to Mexico, where they suddenly turn sharply west and head up to settle in for the winter on 12,000-foot-high peaks, in clusters sometimes so thick that their weight causes the tree branches to break.

There they spend the winter huddled together to stay warm, until in late February they descend onto the Mexican plains. They feed on nectar, lay eggs on milkweed, and begin spreading north once again. The *National Geographic* story was the talk of the planet for quite a while. It seemed impossible: an insect so tiny, migrating sometimes thousands of miles. How did they do it?

I was about to find out that there was more to the story than just pretty wings and marvelous flight behavior. Hoping to learn more, I headed to the West Coast to meet Amelia and her scientist colleagues.

California, it sometimes seems, has been specifically fashioned to try the human soul. Floods. Fires. Landslides. Earthquakes. Deadly droughts. Deadlier avalanches. Massive brush fires raging across thousands of acres. The sudden collapse of whole mountainsides.

If you've never been to California but only learned about its natural disasters in the news, you might well wonder why anyone would live there. At least I myself was wondering this one disappointing February day in 2017, a few months after Amelia's butterfly set out on her journey.

Normally, winter rains arrive in California from Hawaii and points farther west via an atmospheric river of humidity called "the Pineapple Express." It is the Express's express duty to deliver moisture to the state, but in recent years, the weather system has been derelict. Consequently, by 2016 the state had experienced a severe drought emergency. All life had suffered. Vegetation had failed to thrive, which meant that animal life was stressed. Humans had been advised to adopt abstemious behaviors, from declining to wash cars to avoiding toilet-flushing whenever hygienically acceptable.

The following year, the rain gods were apparently trying to make up for past miserliness. Water flooded the state with alarming tenacity. What, I wondered, would happen to the butterflies? The year would turn out to be Northern California's wettest year on record, and statewide the second-wettest, according to the California Department of Water Resources. Rain-related disasters were a dime a dozen. While I was there, about 200,000 people living below the Oroville Dam, 150 miles north of San Francisco—at 770 feet the nation's tallest dam, made of *dirt* (yes, really)—were evacuated one dismal night as chunks of that barrier simply washed away.

How was Amelia's butterfly faring in this weather?

After a journey of only nineteen days, her monarch, tag number A4853, had turned up in North Beach, San Francisco, availing herself of a verbena and lantana feast courtesy of a fourth-story roof deck garden. The apartment's resident, Lisa De Angelis, videoed her, looking only a bit worse for wear. When De Angelis noticed the tiny tag adhering to the monarch's wing, she enlarged the picture and saw the email address and the request that observations be sent to the project's coordinating scientist. The video of the feeding insect soon landed in David James's email box.

By then the butterfly had flown 470 air miles, or a bit less than 25 miles a day. I was surprised to learn this. I'd never thought of butterflies as being so, well, *directed*. This was one focused little life-form. Despite the long flight, James thought A4853 looked in pretty good condition. After a trip of that length, butterflies often have wings that are tattered and torn and lusterless, bedraggled, with maybe a few triangular bites from birds' beaks. Amelia's monarch still looked vibrant.

To find a butterfly that's been tagged is a somewhat rare event, a happy serendipity. Amelia and her mother had actually released twenty-two monarchs, but only this one found its way back into human awareness. Begun in 2012, James's all-volunteer tagging program had released 14,000 monarchs by 2016, of which only 60 had been seen again.

This North Beach sighting was particularly helpful. If tagged butterflies are found, they are usually found on the ground, dead. To find a living tagged monarch was delightful, James said. And that, he thought, was the end of it.

He was wrong. Twenty-three days later, volunteer monitor John Dayton saw her resting on a cypress tree with about 10,000 other butterflies at the seaside Lighthouse Field station in Santa Cruz, south of San Francisco. That was a fun fact, but there was more to come. On November 25, Aleece Townsend of Oregon spotted Amelia's butterfly a few miles away at Natural Bridges State Park.

This seemed a curious choice. Natural Bridges used to be a super-popular overwintering hangout for monarch butterflies, but in recent years only a few thousand generally turned up to overwinter. Were the monarchs dropping in numbers overall? Or had something happened to make the site no longer useful?

Part of a large group of dedicated monarch monitors from the Rogue Valley of southwestern Oregon, Townsend had made the six-and-a-half-hour drive south to Santa Cruz after meeting James years earlier and becoming infected by his enthusiasm. She was already familiar firsthand with the drop in their numbers. "At one time in the Rogue Valley there were thousands of them," she told me. "Now you only see them occasionally." She was happy to report the sighting to James.

But Amelia's monarch still wasn't ready to snuggle up for the winter. On December 30, John Dayton headed over to Moran Lake, a few miles across a cove from Natural Bridges. And there he saw butterfly A4853 roosting in a tree, apparently none the worse for wear. Nothing like this had ever before been reported—four separate sightings of the same insect from four separate sites. Amelia's butterfly, a restless little thing, was upending much received wisdom.

Seven

A PARASOL OF MONARCHS

. . . a rain of golden sequins falls before you.

Robert Michael Pyle

Butterflies have secret lives, and they aren't about to give up their tightly held mysteries easily. Decades ago, the destinations of both populations of North American monarchs, east and west of the Rocky Mountains, were a conundrum. Scientists believed that monarchs west of the high mountain range flew to the California coastline, but there was little hard data to confirm their suspicions. Now, with research from James and other scientists, we know for certain not only what their final destination is, but what paths they might take to get there.

Additionally, because of James's research, we know that our belief that monarchs settled on a tree for the winter and endured a kind of semi-hibernation is false. We even know that some butterflies can live for quite a while and don't necessarily fly away from their roost at the end of the cold season. In the summer of 2019, James announced that his citizen scientists had found a male monarch that had been released ten months earlier in Ashland, Oregon.

At the end of the resting period, this butterfly did not "leave the coastal overwintering area and head inland as all good monarchs are supposed to do. It appears he chose to hang out beachside and spend his old age sunning near the sea," James wrote on his Facebook page.

It's likely that anomalous behavior is endemic to the species. There will always be outliers. Tracking and genetic studies have also shown that the two populations, east and west of the mountains, supposedly separated by the height of those mountains, are genetically the same. "We thought the Rocky Mountains were like a Berlin Wall," entomologist Sarina Jepsen told me. "Now we know that's not true." Exactly how the two groups intermingle to mate is not yet clear.

Butterflies, as a group, have been around for quite a long time, since flowers first evolved. This is not coincidence. Most butterflies *require* flowers. Most moths—the group from which butterflies spring—do not. Consequently, moths appeared long before flowers first appeared. You can see this just by looking at the numbers. There are some 160,000 moth species (more are constantly being discovered), but only about 20,000 species of butterflies, implying that moths have had a much longer time to evolve than have butterflies.

In other words, when flowers evolved, they gradually enslaved some of the moths and turned them into butterflies, who would perform important duties for their flower masters. Flowers turned out to be rather Machiavellian.

"A flower regards butterflies and moths as pollen carriers that can be bribed," say butterfly biologists Daniel Janzen and Winifred Hallwachs in their visually exquisite book *100 Butterflies and Moths*.

It is flowers that power the global drama in our natural world today. Without flowers, butterflies would likely not exist.

Nor, indeed, would we.

I was contemplating the fortuitousness of flowers while riding in a car that February with entomologist Kingston Leong. We were on our way to visit some of his favorite monarch groves. California used to have more than 400 coastal sites where monarchs gathered for the months between October and February. In recent years, about half of these have been abandoned. That might be because the sites changed over time and no longer met the butterflies' needs, or because monarch numbers are dropping, or because monarchs are simply more quixotic and nomadic than we once thought.

In any case, sites currently range from a bit north of San Francisco all the way down the coast to around Los Angeles. They vary widely in the quality of the habitat provided and in the number of butterflies present. Some are used every year, others not.

To see for myself, I had made it down to the central California coast. Showing me a few of the many sites he was watching over, Leong wanted to impress upon me the variety of options from which monarchs might choose.

The morning we met, the rain continued to fall, although delicately. Mist lay over the Pacific so thickly that the planet's largest and most dominant body of water was invisible only a few feet from the shoreline. Looking out from my seaside hotel, all I could see was fog. I put on another layer. I started to sneeze.

First we visited a well-known location, super-popular with the butterfly-tourist crowd: the Pismo Beach Monarch Butterfly Grove. Records show that the insects have gathered here for many, many decades. Leong pointed out that this site of only a few acres was close to perfect. The butterflies were near enough to the Pacific to enjoy its temperature-moderating influence, but just far enough away to avoid being buffeted by violent winds and rains when ocean storms raged. It was far enough south that temperatures were cool at night, but not cold, and far enough north that daytime temperatures did not get hot enough to kill the monarchs.

Pismo Grove sits on a small nugget of Golden State property, mostly dedicated to campsites and beachgoing. The monarchs use only a tiny section. When they are in residence, from about October to about mid-February, volunteer docents present free public talks. They meander along the short paved walkway leading through the small butterfly grove where visitors can look up above to see the resting insects. The docents are available to answer visitors' questions, as well as to ensure that people don't inadvertently step on butterflies that have fallen off their tree branches. This happens rather often.

Thousands of people come here each year to see the butterflies. On the several days that I visited, a steady stream of humanity paraded quietly through the grove, mostly engrossed in craning their necks to see the insects in the branches high overhead. Retired folks with expensive spotting scopes mingled with people in wheelchairs and mothers with newborns. I saw men, women, Germans, Americans, Canadians, Spanish-speaking families, women in headscarves, people carrying umbrellas, people of all skin colors. The bad weather seemed to deter no one.

The grove had clearly become a kind of shrine, an endpoint for butterfly-loving pilgrims. Consequently, Pismo Beach was heavily invested in butterfly tourism. Signs touting the monarchs clustered around the village center like the butterflies clustered on the tree branches. If you came looking for butterflies, at least you would know you were in the right place. Even the local bakery sold cookies in the shape of monarchs, with dark icing mimicking the veins on each orange wing.

Leong parked by the designated grove, on the shoulder of busy Highway 1. It was early in the morning, but the highway was already busy. A stream of eighteen-wheelers rumbled by, along with smaller trucks, motorcycles, and sedans. Horns honked. Brakes squealed. Next to the highway ran a set of train tracks, and next to that was a colony of bungalows packed together on tiny lots. It seemed an odd place to go to commune with an icon of nature. Weren't the butterflies disturbed?

Leong answered that he had brought me here so I could get a sense of the kinds of environments that butterflies can tolerate. Of course, he said, we don't know how well they survive in such places. The noise and pollution may well be impacting them. Recent research suggests that this is indeed the case, but the insects continue to arrive year after year, albeit in ever-decreasing numbers. The seemingly delicate monarchs appear able to coexist with an astonishing array of human behaviors.

Why do they come? Leong and others have found that monarchs overwintering in California prefer a rather specific "microclimate"—a grove of trees with enough branches to allow large numbers of butterflies to settle near the sea, but with protection from strong winds and open to the warming of the sun at both midmorning and midafternoon.

This sounded like a tall order to me, but Leong was about to show me that there were more such sites than I had imagined.

After Pismo Beach, we visited several other small parcels of land, a few acres here and a few acres there, scattered throughout the highly developed region. I had imagined that the monarchs would need vast parcels of land set aside. I was mistaken.

Among the West Coast's butterfly fans, Leong is famous. He fell in love with monarchs decades ago when he visited his first overwintering site. Looking up into the trees as the sun appeared, he saw the insects spread their wings. The experience was almost metaphysical, like looking at a stained-glass window in a cathedral. Leong felt as though he was seeing nature's version of Paris's Notre Dame.

Even after retiring as a professor, he devoted his life to ensuring that monarchs have a future along the California coast: "I'm creating winter resorts for them." He was not being whimsical. When a landowner in the area finds monarchs on his property, Leong is the person to call. Such an event is not unusual. A landowner might well buy monarch-free acreage, only to find it colonized later as trees grow and wind and temperature conditions evolve.

More often these days, though, it's the other way round: acreage that once hosted monarchs ends up abandoned by the insects over time. Some of this may be due to overall population decline, but there are also other factors at play. That's where Leong comes in. If the quality of the site itself is declining, Leong tries to figure out why and tries to improve conditions. Essentially, he figures out how to farm for butterflies under natural conditions.

When landowners contact Leong, he visits the site and creates a management plan that looks far, far into the future. Leong teaches landowners about trees: like other life-forms, trees are not static. Some grow old and fall over. Others must be planted to take their place. Those plantings must be made years ahead of time, since the trees require time to grow. Proper planting requires long-term planning.

But what, exactly, is it that monarchs want? Leong has spent these final years of his own life—"I have another ten years and then I'll be done"—considering that very question. He thinks of this as his legacy to the planet, payment for services rendered.

The first thing he does when he visits a site is to create a wind profile. Overwintering butterflies don't do well in strong winds, which can easily knock them to the ground. So he wants to know where the prevailing winds come from. How strong are they? Are they buffered, either by trees or by topography? Are there older trees that might well fall over in the next several years? If so, should he plant young trees now that might be tall enough in the future to replace the older ones? Where should they be placed?

He also studies the sunlight, which is essential to warm the butterflies after a chilly night. What time do the rays shine through the branches in the morning and in the afternoon? Is the sunlight strong enough—but not too strong—at just about 10:00 a.m. and again at just about 2:00 p.m.? Butterflies, it turns out, are eminently civilized, fluttering around in midmorning and midafternoon when sunlight is optimal. When he studies a grove of trees, Leong tries to make sure that there are no other trees or thick branches obstructing the sunlight at these appropriate times.

A high-end developer once asked Leong to create a monarch grove in the midst of a huge housing development. There had once been a monarch aggregation on the property, but the insects had since nearly disappeared. Only a few hundred were left. Leong began working on that site and picked out a second site he thought the monarchs might use. He wanted to know if he could tempt the butterflies into homesteading the new grove of trees, even though he had no evidence that it had ever been active. It worked. After several years, the butterflies have begun to show up. The housing developer now uses the butterflies as marketing mascots. Images of monarchs with beautifully spread wings are everywhere—even on the walls in the bathrooms.

So it turns out that if you plant it, they will come.

Finally we drove to what was, for me, the most unexpected site of all and my personal all-around favorite—the Morro Bay Golf Course, owned by

San Luis Obispo County Parks and Recreation. The clubhouse parking lot was full, despite the early hour, so we drove a bit farther and parked on a road shoulder. Then we trooped across the course, in between players, who were out in numbers despite the chilly drizzle.

In general, walking through a heavily used golf course is enough to cause hysterics. You're treading on sacred ground. Certainly, staff will soon show up and escort you off the course.

The golfers looked at us quizzically for a second, then nodded in the direction of a grove of tall trees positioned near the peak of a hill with a fantastic view of the ocean.

"Butterflies?" they asked.

"Butterflies," we nodded.

And all was well.

I have never seen a more beautiful setting for a golf course. From the greens, golfers look out over small hills dotted with cypress to the ocean. In the midst of this small-but-busy course, an area is set aside that contains a large number of cypress trees, on which, when we arrived, were hanging thousands of monarch butterflies. Their wings were folded, so that the trees' branches appeared to be covered with dried-up leaves. In the past, there have been as many as 100,000 butterflies here. Last year, there were 24,000. This year, only 17,000.

There are many reasons for this decline, but one specific to the course is that many of the trees buffering the monarchs from the ocean have in recent years been destroyed by wind and storms. The insects are no longer as well protected as they had been. Leong did a wind study and pinpointed several locations where new trees could be planted. He removed a few trees which he felt were interfering with the sunlight the insects needed. He pinpointed another area which he felt with time could become another good overwintering site.

He was farming for butterflies in the middle of a golf course. Again, I was surprised that the butterflies could coexist with humanity's odd behaviors—this time, to wit, the knocking around of hard, tiny, round, white objects. I would have expected the insects to at least prefer something a bit more bucolic, but this course had been active for nearly a century, so the butterflies must have made their peace with flying golf balls.

Still, I was curious about why a golf course would want butterflies. I had a long conversation with the course's supervisor, Josh Heptig, just returned from a national conference on golf course management, where he had received an award for environmental stewardship.

He spoke differently than any golf course manager I'd ever met. Heptig credits his unique point of view to an experience he had years earlier in college. His golf course management class was studying the opening of a new golf course not far from his classroom. Golf course opponents were carrying signs that said, "Birds not birdies." Opponents were invited into the classroom to discuss their concerns. The experience opened his eyes.

He decided to choose both birds *and* birdies. Why not manage for both? Or butterflies. At Morro Bay, following Leong's advice, he had kids from the ages of three to seven plant 80 cypress trees. Then, to celebrate the golf course's fiftieth birthday, they planted another 50 trees.

Not everyone is happy about the trees. You would think the golfers would complain, but if they do, they're fairly low-key with their concerns. They know what they're getting into when they come to the course. The most upset people are local property owners who don't want their ocean views blocked by the newly planted trees. The kids are pretty happy, though. Whenever they see Heptig in the community, they ask if they can come over and water their trees. You can't satisfy all of the people all of the time.

When the butterflies are in residence, from November to February, Heptig leads tourists over his greens to the monarch groves in order to tell them not only about the value of butterflies, but about what he sees as the future role of golf courses in an increasingly crowded world.

"Welcome to the golf course," he begins his tours. "You're here—but did you ever imagine that you would see butterflies on a golf course?"

He sees himself as in sync with Scotland's St. Andrews, which bills itself as the "home of golf," and claims 600-year-old records of golf being played. St. Andrews keeps lists of the birds that use their lands, and has planted special wildflower areas meant to attract pollinators, including several species of butterflies. Even some types of thistles, reviled by gardeners but beloved by many species of butterflies and birds, are encouraged.

Another golf course Heptig manages has won an award for its successful protection of wildlife habitats. He has built raptor perches to attract birds, which will decrease his need to use chemicals to eradicate golf course pests such as moles. He has increased the wildlife on his courses by about a third—a fact of which he is quite proud. "Obviously, the animals are in the area. They just choose to stay now on our golf course."

"We have to have a sustainable business model," he explained, and by "sustainable" he means opening up the course to a variety of activities. Morro Bay is sometimes used for community activities, like footraces. At Heptig's Dairy Creek Golf Course, golfers wind their way through fields of bleating newborn lambs and are watched carefully by hawks enjoying the sights from perches designed especially for them. That golf course is zero-waste: all organic discards, from food waste in the clubhouse to grass clippings, are placed in compost systems, then applied as fertilizer to keep the greens healthy.

So it's not surprising that when he saw the dwindling numbers of butterflies using the Morro Bay overwintering site, Heptig wanted to do something about it. The site is already locally famous. His butterfly tours are well attended. Now Heptig is in the design phase of creating a covered and paved walkway from the clubhouse to the butterfly grove, so that everyone can look at the insects without getting popped in the noggin.

Leong and I talked about these long-term goals as we stood in the chilly air on the Moro Beach course and looked up. With their wings folded, hanging on branches, monarch butterflies look like drab, not-yet-fallen dead leaves—camouflage evolved to signal to troublemakers, "Move on. Nothing to see here," while the butterflies rest.

I was a bit nonplussed. Was this what I myself had migrated thousands of miles, all the way from Cape Cod, to see?

I kept my comments to myself.

We stood quietly for a moment or two. I suggested lunch. It was about two o'clock. I shivered under the bleak sky. I was thinking about a hot bowl of chowder.

Then, as though at the appointed hour, the clouds parted. Rays of light appeared. Blue sky. Sunshine. A brief respite of warmth.

The clusters of dead leaves took flight. The air above us was filled with the colors of orange and black luminescent wings, and with blue skies and puffy white clouds. Below our bluff, the Pacific Ocean glistened.

Overhead, protecting us from the sun, a veritable parasol of monarchs lilted through the air. Buoyant and directionless, full of what we humans would call "joyfulness," the butterflies were soaking up the sun.

Eight

THE HONEYMOON HOTEL

From the mouth of the god of creation each day the sun appeared
and during the winter its rays became butterflies.

Indigenous Mexican saying

Butterflies love their privileged routine. Like royalty, they don't have
to get up until they feel like it, which is, as I have already mentioned, generally around 10:00 a.m.

This adherence to what used to be called "bankers' hours" implies
considerable intelligence, to my way of thinking. I wake up early, but
don't really stir until just about the same time. Which is why I was standing rather grumpily near the parking area at California's Pismo Beach
butterfly site at about 9:30 a.m. the day before I met Leong. I was out
and about way too early and hadn't had my proper thinking time. Additionally, the weather was—surprise, surprise—dismal and dank.

Monarchs, too, dislike rain. They stay huddled on their branches
in their dead-leaf poses. For butterfly tourists, there's not much to see.
Fortunately for me, upon arriving and finding little excitement, I learned
that my early start would, after all, be rewarded. Even if the butterflies
were reluctant to get up, my workday would be salvaged. A public talk
was scheduled for 10:00 a.m. Even better, a devoted volunteer docent,
one of scores I would meet while researching this book, promised to
reveal the secret love lives of the Pismo Beach monarchs.

I adore secrets.

I settled right down on one of the park benches.

People began to gather, despite the weather. There were well over fifty humans present, bundled up, warmly hatted and carrying umbrellas to ward off the intermittent showers. We were eager to learn more about the splendor of monarchs. The small crowd included a multitude of children, quite a few of them just above toddler age. This fact would later turn out to be important: the volunteer, an amiable woman who looked as though she'd just retired from a teaching job, had matters of considerable delicacy she needed to discuss.

She began by explaining the basics. Female monarchs lay eggs about the size of a pinhead on milkweed plants—and *only* on milkweed plants. They deposit the eggs on the underside of a leaf, usually one egg to a leaf. Monarchs are not the only insects that use milkweed. Well over a hundred other species do, implying that at one time milkweed plants were much more common than they are today. Still, the competition isn't quite as tough as it might seem, as not all of them use the same parts of the plant as the monarchs.

After three to five days, depending on the weather and the temperature and the time of year (more about this later), a miniscule caterpillar emerges. The emergent life-form is so small I looked directly at one but did not see it.

The caterpillar spends the next nine to sixteen days desperately eating milkweed, starting with the leaf on which the egg was laid. The caterpillar must eat milkweed—and *only* milkweed. Poor thing. The creature has no choice in the matter; no other plant but milkweed will do. For the monarch caterpillar, the intensity of survival begins immediately upon emerging from the shell.

After first eating the egg's shell, which is full of nutrients, he takes a long drink, like a "cat drinking milk," according to the entomologist Dame Miriam Rothschild. In fact, he goes on a milkweed-sap bender, drowning himself in the stuff—sometimes literally, as the ecologist Anurag Agrawal has documented.

Most of us, if we have had any kind of experience of nature as children, are familiar with the ickiness of milkweed sap. Tear open a milkweed leaf, and out the rubbery goo would "spew," monarch biologist Lincoln Brower told me. And then it would dry, and then your fingers would be stuck together. It was fun as a kid to pretend that your fingers had been captured by this mysteriously strong and compelling blob-like substance, which would stay gummy and mucilaginous and annoying and never come off your hands until you washed it off painstakingly in a nearby stream.

As adults, we would learn that this material is called *latex*. Latex is by no means rare. Roughly ten percent of all plant species have evolved to use latex. Latex from rubber trees makes our automobile tires. Synthetic rubber has been developed, but is not as durable as natural latex. There is nothing else quite like it on the planet.

Milkweed latex is all-around nasty stuff, full of poisons. The famed monarch researcher Lincoln Brower once tasted some: "It just about knocked me over. It was so foul-tasting. I drooled and I almost vomited."

Interesting, I thought. I myself am not in the habit of putting strange materials in my mouth. But field scientists seem to display a certain machismo, which they can then crow about over beers. Even Charles Darwin succumbed to this derring-do: "One day, on tearing off some old bark, I saw two rare beetles and seized one in each hand; then I saw a third and new kind, which I could not bear to lose, so that I popped the one which I held in my right hand into my mouth. Alas it ejected some intensely acrid fluid, which burnt my tongue. . . ."

Most scientists survive these experiments. . . .

Another butterfly expert, perhaps the most widely known one of the twentieth century, Vladimir Nabokov, wrote that once when he was in Vermont, to test the similarity of monarchs and another butterfly known as a viceroy, he tasted both and found them similarly "vile." Nabokov's remark received considerable publicity, primarily because he was massively famous as the author of the shocking *Lolita*, about the obsession of a middle-aged man for a young girl. (This was in the 1950s, when writing about such things was taboo.)

The fact that the just-emerged caterpillar must eat milkweed is a strange irony: the caterpillar's first drink might well be its last, since the latex that glued your fingers together when you were a kid can also glue caterpillar jaws shut. Hence, death by starvation. Agrawal suggests that roughly 60 percent of caterpillars die because of this first meal. That's a pretty high casualty rate. If their jaws are not overwhelmed, then the animals may well die because of the rather mundane fact that their feet get stuck.

Sometimes, to lessen the danger, a caterpillar will succeed in separating the leaf from the plant by chewing at the joint between the leaf stalk and the main stalk of the plant. This makes his job less onerous, since the latex spewing out of the leaf will do so with less pressure, and the caterpillar can partake of the nourishment in smaller sip-size swallows. Or sometimes he will even carefully chew a circle in the leaf, then eat from inside the circle, thereby vastly reducing the pressure of the gushing latex. Mostly, though, he will just begin chomping away and hope for the best.

His compulsion to eat of the leaf of the milkweed plant and to drink the bitter latex seems terribly cruel, the stuff of Greek tragedy: Condemned to embrace the source of our own demise. Fatal attraction. There's more to this dire truth than just the stickiness of latex and the fact that it tastes atrociously bitter.

Latex is deadly. Another irony: the more latex a caterpillar eats, the more reduced that caterpillar's overall growth, *but* the better protected that same caterpillar will be from birds and other predators. If, that is, the caterpillar survives. Many die from the poison they are required to ingest.

This is perverse.

It's also essential: overall, predatory birds cut caterpillar numbers in half.

That latex is poison has been known by humans for centuries. Romans extracted it from plants and used it to assassinate enemies. It is something like digitalis, affecting the heart and nervous system of all animals.

So here's something to chew over, perhaps the ultimate perversion:

the caterpillar drinks the latex precisely *because* it's poisonous. That which does not kill us makes us stronger. If it survives the poisoning, it's way ahead in the life-and-death match it will be playing for its entire existence.

The caterpillar can store the toxin away in several specific locations in its body. When predators try to eat the caterpillar, they get mouthfuls of noxious, repulsive poison. Birds find this discouraging. Often their response is to regurgitate. Predators learn quickly. Most never again try to eat a monarch. Or, as Darwin and Bates showed, other butterflies that just *look like* monarchs. These stored toxins have lasting power. When the caterpillar transforms into a butterfly, the stored toxins remain, still able to deter predators.

We ourselves are a lot less tolerant of this toxin than are the monarchs—as witness Brower's drooling response—but we, too, can derive benefits from consuming it in super-small amounts. "There is often a fine line between poison and medicine," Agrawal writes in *Monarchs and Milkweed*, his in-depth discussion of the conundrum. The medicine many people take for a variety of heart problems is closely related to milkweed toxin. It may even someday be used as a cancer drug. On the other hand, consume too much of it and you will go into cardiac arrest.

Monarchs have evolved all kinds of strategies to get around the terrible conundrum so central to their existence. However, milkweed has countered with new and even more devious ways to get around their monarch problem. Some milkweed species grow bristles on their leaves to make the caterpillars' job even more formidable, for example. In that case, the caterpillar first becomes a living, breathing mowing machine, chewing away the spines before eating. This to me seems incredible: How does this tiny, temporary creature, only one small phase of a complicated life cycle, "know" how to do this?

Monarchs want to eat milkweeds and milkweeds don't want to be eaten. Agrawal and others characterize this as an "arms race," a tit-for-tat kind of thing, a card game for big bucks, an I'll-see-you-and-I'll-raise-the-stakes endeavor. But still others suggest that the metaphor may be reflective more of the workings of human society than of insects and plants.

The simple word "interaction," suggesting an interplay of many different ingredients, is less loaded and more accurate. Entomologist Michael Engel uses the phrase "evolutionary back-and-forth." After all, Engel told me: "The ecological rise of flowering plants was partly fueled by their marriage to insects, and many groups of insects owe *their* own success to their floral hosts."

I asked Agrawal why some newly emerged caterpillars succeed while the majority don't get past their first day of life.

"In nature, an oak tree produces a million acorns over a lifetime," he answered. "Why do some survive? Some of them survive by chance, and some of them survive because they have the right traits. The same is true in this example. But the more scientific answer is that monarchs and milkweed don't live in isolation from each other. They are each trying to do better at their own jobs, through natural selection. Maybe if you prevented the monarch from evolving over a thousand years, but the milkweed continued to evolve, all the monarchs would die out."

But of course, that would never happen.

All life must evolve.

Change, after all, is the very nature of existence.

Those are the tectonic facts of life.

When the caterpillar is about two inches long, the Pismo Beach docent continued, it begins the process of turning into a butterfly. It sheds its caterpillar covering, retreats from a dangerous world into its cubbyhole-like chrysalis, and gets the job of transformation done in relative safety, out of sight of predators. This generally takes only a few days.

When the butterfly emerges, she will be an *imago*—a fully formed adult—and will wear the fabulous colors which so attract the human eye. But the monarch colors which mesmerize us have a purpose other than dazzling: they are skull-and-crossbones warning signs. Eat me at your own risk. Monarchs are not the only insects that eat milkweed and warn off predators with those intense orange colors. Milkweed beetles and milkweed bugs also have threatening orange coloration that cause them to stand out loudly on the green plant leaves.

Nature does that sometimes. Usually, prey animals wear the colors of camouflage, colors that make the animal difficult to see. The white dapplings on a fawn's rump mimic spots of sunlight on dried grasses. Zebra stripes hide the prey animal from prowling lions. But if the animal has a unique defense, the strategy instead might be to use colors to make the animal stand out and be memorable. Skunks, for example, are startlingly black-and-white. Their coloring is a kind of braggadocio: I have nothing to be afraid of—but *you* do. My black-and-white border collie with his intensely dark, menacing eyebrows uses the same strategy: he *wants* the sheep to notice him.

Spider wasps also use a rich orange color as a caution light. This insect's sting is painful, even dangerous. But the insect would rather not have to push things that far. So, in strategic spots, this black insect wears orange: orange wings, orange antennae, orange bands around the abdomen. This venomous creature might as well be wearing a jacket that says "I really don't care. Do U?"

The same is true of monarchs. That glistening bright orange on the wings is a notice to stay away. *Warning:* If you don't, you won't like what happens. Newly emerged monarch caterpillars, who have not yet accumulated toxin, are not orange, but are instead almost translucent. Defenseless, they need to hide. But as the caterpillar grows and accumulates more toxin, it can afford to take on brighter colors. By then, standing out like a neon sign is a defense. With black, yellow, and white bands, it can be bold; it now has enough stored toxin to be effective. Go ahead, the colors say, make my day. Take a bite. You won't come back for more.

That chilly February morning on Pismo Beach, we all stood listening as the docent talked. From time to time the sun managed to send rays through the rain clouds.

When the transformation process inside the chrysalis is complete, the volunteer explained to the rapt children and adults, a butterfly emerges. In the chrysalis, the wings have been folded, so the emerged butterfly must first take an hour or so to pump fluid into her wings. They straighten out and become more firm. Then the butterfly flies off in search of nectar and love.

Then, the volunteer said, the circle of life begins again.

"And there they go, right on time," she exclaimed.

Slowly, one by one, then more and more, the monarchs left their perches on the tree branches and began flying around. I assumed they would benignly begin looking for flowers with nectar.

But something else was happening. Some seemed to be chasing others. Those being chased were doing acrobatics, twirling upward, flipping around in the air, trying to avoid the chasers.

How odd, I thought: they're playing a game of tag.

Then one of the chasers caught up with the chasee. He tried to grab on to her. She managed to avoid him. He grabbed her again. Then, unable to clasp on to her securely, he knocked her to the ground. He followed her and began trying to pin her to down. She struggled. Eventually, his hold on her was complete. He flew up into the air, carrying her with him.

"And then," said the kindly retired schoolteacher, "they fly off to the Honeymoon Hotel."

She left it at that.

Not everyone who describes monarch sex uses such gentle language. Indeed, some lepidopterists who choose to discuss the subject are even a bit indignant.

"The Monarch butterfly could well be designated nature's prime example of the male chauvinistic pig," Dame Miriam Rothschild, niece of the collector Lord Walter Rothschild, once wrote. "The other members of this genus bemuse and subdue their females by means of a sophisticated aphrodisiac—a love dust metabolized from a plant precursor and shaken over her like a golden snow flurry during courtship. The male Monarch on the other hand, dispenses with these refinements and, more often than not, knocks down his female, and while in a half dazed state, takes her by force. Her antennae may be bent double in the process, her legs crumpled beneath her body and her wings sadly mutilated."

Written in 1978, her essay was titled "Hell's Angels."

When I read this, I was taken aback. Then I was grateful that the Pismo Beach docent hadn't been this open and honest. Not only did kids not need to know this dreadful fact, but I wasn't sure that even *I* needed to know this. Given a choice, I generally prefer to live in a complacent world of sweet illusion.

Of course, interpretation is everything. If Rothschild was more than a bit anthropomorphic in her description of monarch courtship, it's because, like monarchs, she could afford to be bold. Like adult monarchs, she was quite well protected. As Dame Miriam Louisa Rothschild, a scion of *the* Rothschild family of banking fame, she could afford to be pretty much anything she wanted to be, and what she wanted to be was an entomologist. Even though she had no formal education, until her death in 2005, at the age of 96, she was considered one of the world's foremost experts on monarch butterflies.

She also knew a lot about fleas. Her father, Charles, a flea lover since childhood, had collected more than 260,000 of them. If you have the compulsion to collect, and if money is no object, collecting hundreds of thousands of miniscule insects is not necessarily beyond your wildest dreams—if indeed those *are* your wildest dreams.

This was more than just a silly rich man's hobby, like uselessly collecting Rolls-Royces or diamonds or palaces or something. Charles's pack rat compulsion was coupled with a genuine scientific talent that profoundly benefited humanity. In 1903 he discovered a species of flea, *Xenopsylla cheopis*, that lived on rats but frequently jumped onto and bit humans. This flea, Charles found, was the vector for the implacable waves of bubonic plague that had repeatedly washed over human civilization since at least the sixth century. Because of Charles's discovery, we now take fleas very seriously, and bubonic plagues are considerably more rare.

Miriam, too, made groundbreaking discoveries about fleas—"It isn't everyone that has a great love of fleas," she once said, "but I have"—including the mechanism by which fleas make their fabulous jumps. "We found that these fleas could jump 30,000 times without stopping," she later said. "It was really rather a lot. . . . The acceleration actually turned out to be 140g and that was twenty times the acceleration of a moon rocket reentering the earth's atmosphere."

I thought about this: it takes a lot of dedication to fleas to figure out such things. She was right: not everyone feels about fleas the way she felt. Miriam Rothschild is one of my favorite characters in butterfly science. Forbidden to attend school as a child (girls don't need education), she was almost entirely self-taught and became a serious scholar. As an adult, she was associated with several well-respected universities and research institutions. She organized the world's first international flea conference, at which scientists were entertained both by Ginger Baker of the rock band Cream and by the refined performances of a string quartet. By the time she died in 2005, she had published more than 200 papers in academic journals and been elected a Fellow of the Royal Society. She is also said to have invented the seat belt.

Walter Rothschild set an excellent example for her. He lived life on his own terms and once drove a carriage pulled by zebras to Buckingham Palace, but didn't stay long because he knew there would be trouble if one of his draft zebras, none too tame despite their apparent willingness to travel London's cobblestone streets, bit one of the royal children.

Miriam shared his tendency to tease the social elite. She often wore tentlike purple dresses topped off by purple scarves. She wrote many books, wore white rubber boots to Buckingham Palace underneath her evening gown, spoke out for what we now call gay rights, and encouraged the wildflower gardening craze still spreading today. She donated a large amount of money to schizophrenia research and pioneered art therapy techniques.

Her scientific interests were varied, but usually involved pursuing a question about which she herself had become curious. One day, for example, she wondered about the relationship between tiger moths, ear mites, and bats. A particular species of mite preyed on tiger moths, which were in turn preyed upon by bats. Miriam noticed that these ear mites infested only one ear of the moth. The other ear always remained mite-free. Miriam hypothesized that the mites left one ear free so that their free ride—the tiger moth—could hear an approaching bat and escape.

"Once . . . one mite gets in, other mites follow. But they always go into the same ear. The moths never have both ears filled with mites. And you see these mites fighting one another. And copulating with one another and so. All within one ear. And nobody understood this. Nobody knew why they only go into one ear. Seemed very odd," she once told a television interviewer.

She found that the first mite left a trail into the moth's ear, which subsequent mites followed.

"Because obviously, the mites don't want to get eaten by bats either. So this is a protective device. And I think it's a very amusing one," she said.

"I must say, I find everything interesting," she later said.

And she did.

I imagine her thriving in the Italian Renaissance and having intense conversations with the likes of Leonardo da Vinci. She would easily have held her own.

One of the things she found interesting was the immunity of monarch butterflies to predators. Although monarchs are not endemic to Britain, she had read a lot about them and had had some samples brought to her. Asked to pick her own personal seven wonders of the world, she once explained to an interviewer: "I chose the monarch butterfly. . . . Because butterflies altogether, you know, are very smart.

"They *smell* very strongly," she said, meaning not that monarchs are good at detecting aromas (which they are), but that the insects themselves emit strong odors.

I assumed she meant that the toxins had aromas, but Anurag Agrawal set me straight: many insects have repellent smells, but the monarch's aroma is not due to milkweed toxins, as those toxins are too heavy to become airborne like volatiles.

Then what causes the smell, I wondered?

"Unknown," was his answer. "This would be ripe for future study."

In a video interview, Rothschild once held up a bluish-green chrysalis spotted with golden dots, containing a monarch caterpillar transforming into a butterfly.

"That little sweetie," she said. Then she explained the groundbreaking science in which she had been involved, along with three other researchers—the American scientist Lincoln Brower, the Swiss Nobel Prize–winning chemist Tadeusz Reichstein, and the British scientist John Parsons.

Since the nineteenth century, observers had noticed that monarch butterflies were free from birds that preyed on many other butterflies. Some suggested that this was because the monarchs were distasteful, but no one had proven this. Additionally, there was the question of *why* the monarchs were distasteful. Were the butterflies making their own toxins? Or were they consuming the toxins in something they ate?

The answer to this question, common knowledge to us now, was not so easily teased out. Nineteenth-century observations had shown that monarchs ate only milkweed, that milkweed sometimes killed caterpillars, and that milkweed itself was both bitter and poisonous. It stood to reason that something in the milkweed was likely protecting the monarchs. Evolution, still a controversial science at the turn of the twentieth century, seemed unable to explain the obvious connection. How could two separate organisms—the plant and the butterfly—have evolved such an intimate connection?

The mystery lay in limbo until the 1960s, until Dame Rothschild, the Perle Mesta hostess-with-the-mostest of international biology, determined to solve the puzzle. She did this by inviting the relevant scientists to lunch at her estate. Much fun was had by all. Then letters zoomed back and forth across the Atlantic. Americans Lincoln Brower and Jane Van Zandt Brower spent some time at Oxford and discussed the issues with Rothschild and others. A series of groundbreaking experiments unfolded.

The Browers took the lead in research that worked directly with the butterflies themselves. First they showed that blue jays regurgitated monarch butterflies, if induced to eat them. The February 1969 *Scientific American* issue that discussed the groundbreaking research showed a colorful front-cover illustration of a bemused-looking blue jay trying to decide which of two butterflies to eat.

Next, Lincoln Brower reared a strain of monarch caterpillars that could eat cabbage rather than milkweed. "If you had asked me if that were possible, I would have said no," Agrawal told me, by way of praising the import of Brower's achievement.

This is how he achieved this seemingly impossible task: He took newly emerged caterpillars and set them down on cabbage. Then he left them there. Almost all starved to death.

Remarkably, though, a few managed to survive. To the end of his life, he was enormously proud of this achievement, he told me only weeks before he died in 2018, as it showed just how flexible a species can be.

Brower bred those to each other, then set the resulting caterpillars on cabbage. After repeating this numerous times, he ended up with a small number of monarch caterpillars that not only could survive by eating cabbage, but—and this is important—were *not toxic*. When Brower fed these to predators, they did not regurgitate the insects. Thus, Brower proved that the toxin in the adult monarch derived from the plant eaten by the animal as a caterpillar. (This small example, by the way, is one simple case of how nature handles the need to evolve when changing climates change the kinds of plants available to animal life. If all insects die in an area except for the few that can manage to consume a new plant, they are likely to find each other, mate, and reproduce. Eventually, you may have a new species. One the other hand, all you may get is yet another extinction.)

"We really kind of broke it open with that cabbage experiment," Brower told me. The team was well on its way to founding a new branch of biology: *chemical ecology*, or the study of chemistry as the "language" of interspecies communication.

I asked him if pursuing such a risky endeavor was daunting.

"There was a lot of luck involved in that experiment," he answered, "but there usually is in experiments like that."

I asked him what he meant.

"The experiment was a long shot," he answered. "We put hundreds of monarch caterpillars on cabbage and one survived and could be bred, allowing us to select a stock that could digest the cabbage leaves without killing or starving the caterpillars."

The next task was to learn specifically what the poisoning substance in the milkweed was. Rothschild had enlisted the help of the Swiss chemist and Nobelist Tadeusz Reichstein. Since monarchs are not native to Europe, Brower obligingly sent some to the overseas chemists. Those researchers isolated the poisonous compound from the butterflies. And then they found the identical toxin in milkweed.

That clinched it. Monarchs and milkweed were closely intertwined in a relationship mediated by a specific chemical, a relationship that at one time would have seemed to many people to be pure magic, given that the relationship is based on a currency invisible to us, although quite obvious to the butterfly. We take for granted today that chemistry is the currency of evolution, but in the middle of the twentieth century, this was breaking news that was, to some people, almost unbelievable. Brower went on to become the first president of the International Society of Chemical Ecology.

Miriam Rothschild went on to win eight honorary doctorates, including from Oxford and Cambridge, and to become a Fellow of the Royal Society.

And although she continued to love fleas and butterflies, and remained extremely fond of female monarchs, she never changed her opinion of the male of the species.

"He is a thug," she once wrote.

And she meant it.

Nine

SCABLANDS

The Monarch butterfly is, in my opinion, the most interesting insect in the world.

Miriam Rothschild, The Butterfly Gardener

The western third of the state of Washington is little more than an agglomeration of bits and pieces of the northwest-moving Juan de Fuca tectonic plate. Chips off the old block, if you will, attach themselves to the continental plate and thus does the state of Washington grow. It is as though Oregon's coastal range is being "shoved unceremoniously into Seattle's underbelly," writes the geologist Ellen Bishop.

This makes for some interesting weather. Fed by rivers of rain flowing in off the ocean, western Washington is so wet that, during the long winters of drizzles and downpours, some people simply go mad. Seattle's Aurora Bridge is infamous for its high number of winter suicide jumps. By April 2017, a few months after I first visited Pismo Beach, Seattle had endured 45 inches of rain in six months—the most rain recorded since documentation began at the end of the nineteenth century. Compare that to the nearby Olympic Peninsula, slightly to the west, where rainfall totals topped 100 inches.

It turns out, though, that this occurs only *west* of the Cascades, a 700-mile-long mountain range that's been growing for almost 40 million years. *East* is another story entirely. Washington is known as the

Evergreen State, but on this side of the mountains, that nickname is just a mean-spirited joke. The eastern two-thirds of Washington is hot, dry, forbidding and foreboding. Ancient dunes of silt and sand cover large areas. Where people west of the mountains yearn to see the sun for just a few minutes, east of the mountains people yearn for escape. The sun bakes the land in many places into a kind of impenetrable pavement. On the bleak dark basalt cliffs even lichens have a hard time hanging on. The landscape makes you feel thirsty just looking at it.

You need an airplane to understand the scope here. There are old rocks, sometimes quite old. In a few places near Canada and Idaho, you can touch rocks that existed when the region was part of Kenorland, a 2.5-billion-year-old supercontinent.

This is all supposed to be *secret* history. Our planet has traveled a long way since then, and in any kind of reasonable, normal locale, those rocks would by now be hidden by 2.5 billion years of detritus—sediments and plant fossils and dinosaur bones. Stuff like that. We wouldn't have to be continuously confronted by the cruelties of time.

But the Pacific rains don't make it to this region. Reigned over by Mt. Rainier, at 14,000 feet the most glaciated peak in the continental United States, the Cascades squeeze every possible bit of moisture out of the storms. East of those mountains, Washington State is pizza-oven hot. Threatening. Stark. Stern. Desertified. Dry as dust. The whole area has its own name: the Scablands. The moniker fits.

It was late August. I had come out to meet up with David James, the scientist running the monarch-monitoring program. We met in the town of Yakima, home of thousands upon thousands of apple trees. Washington apples are famous. In the advertisements, pictures of these pleasant orchards reminded me of the Vermont orchards where I picked apples for money as a college kid. Turns out Washington State "orchards" are quite different. No gentle hills. No soft greens.

He warned me that the day's temperature was going to top out at 100 degrees. This didn't surprise me because that's what the temperature had been the day before. I'd weathered that day by floating down the Yakima River looking for trout, and I was ready for this day, too, dressed in a loose white blouse. Unfortunately, though, I also had to wear heavy boots and protective jeans, since we would be tramping through brush. Snake territory.

We were headed for the Lower Crab Creek wildlife area. The creek is all that remains of a once-gargantuan lake. To get there we would drive to the white bluffs of the Hanford Reach National Monument—once home to the Hanford nuclear reactor, our planet's first large-scale nuclear reactor and currently the site of the nation's largest nuclear cleanup effort. Walking trails in this monument, now open to the public, involve dunes every bit as daunting and searingly hot as those I'd found in the Sahara.

We followed the Columbia River for a few short minutes, then turned onto a side road. We passed by the steep north slope of the soot-colored Saddle Mountains. When we got out of the car, I realized that there was no humidity. Even my eyeballs felt parched. Life burned by the sun.

There were monarchs here?

All they need is nectar, water, and shelter from the sun and wind, James explained. In fact, the first few generations of monarchs over that summer had done particularly well. Despite appearances, there had been "lots" of rain (his word; definitely not mine) over the winter and well into the spring. The milkweed had flourished. Although the place was parched now in late August, he said, monarch fertility over the summer had been high because the water levels had remained high.

I asked about the heat.

It does affect them, he admitted, particularly if it lasts over long periods of time. In 2015, he said, there were weeks of temperatures that reached as high as 115 degrees. The butterflies had suffered. At such high temperatures, the butterfly's development slows down. The insects are more vulnerable to predators. Despite the heat, James's long-term studies suggest, the insects that come here may be long-term residents and are not just passing through. When they arrive at this site during their spring migration, they seem to stay and breed, rather than moving on. That, he believes, means that they must be finding everything they need right here in this tiny oasis-like area.

James is an expert in biological pest control who specializes in advising vineyards. His monarch conservation work is done entirely as a volunteer, paid for out of his own pocket. "Beauty with Benefits" is his theme during his day job, as he tries to convince vineyard managers that they can save money by cutting down on pesticide use while planting native wildflower gardens. The idea is to lure in beneficial insects to help control the ones that destroy the grapevines. He's an advocate, not surprisingly, of planting milkweed around vineyards to attract a variety of beneficial insects, such as bees, that will help limit more harmful insect species.

And as a happy side effect, you'll get more butterflies, he likes to point out. Butterflies don't discourage harmful insects, he admits, but they do wonders for the human psyche. No matter what he is talking about, butterflies seem to find their way into James's conversation. It's been this way since he was eight years old and found a caterpillar in his backyard in England. He reared and released it, and from then on he intended to become a naturalist. By 1970, while still a child, he was already publishing articles in local papers encouraging people to plant stinging nettles in their English gardens. (Many butterfly species love them.)

Like the monarchs he loves, James is a migrant. From Manchester he traveled to Australia, where he earned his doctorate studying the populations of monarchs thriving on the island continent. The species, not native to Australia, somehow made its way across the Pacific and was first noted in Sydney in 1871, where it has been quite common ever since. The popular name there is not "monarch" but "wanderer."

"The butterflies got to Australia, we think, on their own, by crossing the Pacific in a gradual spread from island to island. Once they got to Australia, they adapted to the conditions there. I saw that firsthand."

This sounded to me like a long way to fly.

"It can happen," he replied. "Every year one or two monarchs from North America turn up in England." If they can be carried across the

Atlantic by wayward winds, they may well have been carried—or flown—all the way to Australia by island-hopping. That's the theory, anyway.

When they got there, they must have, like the Crab Creek monarchs, found what they needed. Monarchs live in many places in Australia. As noted before, some of the populations are migratory; some are not. In Australia, as in North America, monarchs living in seasonal areas where temperatures become colder in winter migrate to safer climates elsewhere on the continent. They also tend to favor overwintering sites in the same kinds of habitats studied by Kingston Leong, who had first taken me to the Pacific Grove overwintering spot, and others. James published the first study ever done on overwintering monarchs in Australia, in the Sydney area.

"They do migrate in Australia, but over much shorter distances. The need for migration there isn't as great. The decision to migrate is much more flexible and depends on the weather and other environmental conditions that they experience when they emerge. If they experience a period of warm, sunny weather they don't need to migrate. They do both things. They are more flexible."

Later that year, he would find evidence of just how flexible they can be.

Most of what I had read seemed to indicate that monarch butterflies were dominated by innate behaviors, but James disagrees.

He said: "The more you look at them, the more complex they are."

From Australia James migrated to Yakima in 1999 and immediately began looking for monarchs. They are not the only butterfly he cares about. He is the coauthor of an acclaimed book that examines in great detail the life cycle of all 158 butterflies that occur in the American Northwest—the first-ever such book. David Attenborough called this work "magisterial" and has a copy on his bookshelf.

When we arrived at his research site, we parked in an asphalt parking lot beneath the basalt cliffs. James wasn't sure if there would be many monarchs around. Perhaps they had all left. Or perhaps the heat had killed them off. It was the end of August 2017. In past years many would already have started their migrations. But earlier in the summer, numbers had been promising. Perhaps, he suggested, there would still be a few hanging around.

This tiny nugget of wetlands had benefited from the "deluge" of rains that had hit the region over the past winter. I use the word with irony. In January, the region averages about an inch of precipitation. But this past January, while Seattle and the Olympic Peninsula suffered their inundations, the people of Yakima celebrated with twice as much moisture as usual. Instead of the typical one inch, they received two. As a result, even now, in late August, the water table was still comparatively high. A few milkweeds still bloomed.

We walked through the brush. The springtime grasses had dried to a golden-brown, but several plant species still bloomed, including purple loosestrife and Russian olive. Both species are regionally invasive, but James believes that they provide the foundation for a unique mini-ecosystem, one which explains why Crab Creek is the Promised Land in the desert for so much wildlife.

Purple loosestrife is so invasive that some states have outlawed its planting. The garden varieties were not supposed to reproduce, but they met up with natural varieties, hit it off, and cross-pollinated. Native plants are driven out by these hardy flowers, but the truth is: butterflies love them. Loosestrife is enjoyed by swallowtail butterflies, sulfurs, cabbage whites, lots of little blues, and, of course, monarchs. The plant lines the banks of the Columbia River, only a few minutes away as the butterfly flies, and may be the reason why butterflies can come all this way. Monarchs likely follow the river on their fall migration, feeding on loosestrife, which enables them to get a good start on their way to California.

The thickets of Russian olive trees, James believes, are also important. They provide shade from the summer heat and hiding places for females that don't want to be bothered by males. He has found eggs on the leaves of milkweed plants growing under the tree branches.

"I don't know if we'll find anything here this late in the summer," he said.

Almost immediately, he spotted a female. His butterfly net, almost twice as long as he is tall, whipped out faster than a cowboy's lasso.

She was caught.

Carefully removing her from the net, he spread her wings out in one hand, gently holding her by the head and thorax with his forefinger.

"Never by the abdomen," he warned. "That's where the eggs are."

We could see that she had lost a few chunks of wing. Her colors were rather dull. She had clearly shed many scales. This, James explained, was not at all unusual. It merely showed that she had been around the block a few times.

"She won't be going to California," he said.

He carefully scooped her up and put her in his collecting kit, inside a plastic container with mesh on the top to circulate the air to keep her cooler.

I asked if she needed air inside the container in order to breathe. He explained that while she did need oxygen, she used so little that even if the top had not been made of mesh, it would have been a long time before the oxygen ran out.

"She'd be more likely to starve to death first," he said.

That, however, was not going to happen. It's a common myth that after a female lays eggs, she dies. In fact, she can breed and lay eggs for as long as she lives. This female, despite her age, still had eggs inside her. James wanted to bring her home. He was fully supplied with nectar plants and with milkweed. The eggs she laid would become, in his care, a healthy next generation that could be tagged and then released out into the universe, just as Amelia's butterfly had been.

James has tagging projects going on all over the Pacific Northwest. He's a Pied Piper, drawing in people with the soft sound of his voice, with his enthusiasm for all things natural. Only begun in 2012, his tagging project is now famous west of the Rockies. He's got crews down in southern Oregon, crews near the California border, and crews over in Idaho. He's got a few prison programs going, too. Over at Walla Walla, where murderers are in for the long run, he sends eggs on milkweed leaves to be looked after. The prisoners bring the insects along all the way into the flying stage, and then either tag and release them themselves or hand them over to James for release at one of his many regional events.

"The crux of this research is the rearing," he once explained. "Inmates have been shown to be very good at rearing creatures. When you rear large quantities of caterpillars, you get disease problems if you don't pay attention to the hygiene."

Rearing caterpillars is labor intensive. First, they need to be fed on fresh leaves. As they get larger they produce copious amounts of frass (manure). The frass must be discarded so that bacteria don't infect the insects. Caterpillars need to be moved as they grow larger. Chrysalises must be separated, since they are vulnerable to predation by other caterpillars. James considers Walla Walla a golden opportunity: inmates have a lot of time on their hands. The prisoners like the program. When one group was asked whether they would prefer to work with kittens or monarchs, the prisoners overwhelmingly chose the butterflies. They call themselves "butterfly wranglers." This, for James, is the real "Butterfly Effect"—the ability of butterflies to connect people, even incarcerated people, to their roots in the natural world.

It's ironic that an invasive tree like the Russian olive, which all over the American West is pushing out cottonwood and willow trees, may be nurturing monarchs here at Crab Creek. With just the smallest amount of surface water Russian olives can create an extensive dense grove of interlocked trunks and branches that are nearly impenetrable to many animals. We ourselves were able to walk up to these groves, but would have been hard-pressed to follow anything inside.

"There's one. A nice fresh female."

Swing. Caught.

Looks easy. It's not.

To tag a butterfly, you need to close up her wings. You will use the wing's underside because the tag is easier to spot when the butterflies are hanging from branches with folded-up wings. Butterfly wings have veins that form patterns with specific interior shapes. Each of these shapes is called a "cell," which designates a specific area of the wing surface. On the underside of one wing, you place the adhesive side of the tag in the discal cell of the hind wing, or what is commonly called the "mitten" cell of the wing, because it is shaped like a mitten.

"There's another one."

He takes a second look.

"It's an old one," he says as the butterfly departs on the wing.

"How can you tell?"

"It's not as brightly colored. It looks washed out."

Butterflies shed their scales quite easily. If you're not careful when you handle one, you'll find your fingers covered with a pixie dust–like material that turns out to be butterfly scales. People who handle butterflies frequently have to use masks, because the miniscule scales can enter human lungs and cause severe breathing problems.

We caught a gravid female.

I asked: "How do you keep from hurting them when you catch them?"

"They're robust," he answered.

I looked quizzical. "Robust" and "butterfly" did not seem to me to belong in the same sentence.

"Have you ever held a butterfly?"

I hadn't.

He handed her over to me.

"Their thorax is strong. You have to press really hard to do any damage."

He suggested I press. Reluctantly, I followed through with a tenuous increase in pressure. My vision of butterflies as delicate and ephemeral, dreamlike beings that could vanish like Tinkerbell was challenged.

He was right. The exoskeleton was harder than I'd expected. I'd thought I'd be holding a wraith, but instead I was grasping a substantial living being.

"Poor thing," I said, as she was packed away in James's collecting gear.

"She's going to a rest home," he answered. "She's going to be eternal. She's going to lay eggs. She's not going to be interfered with by the males anymore and she's going to have plenty of nectar to feed on."

What more could an aging female monarch want?

When the heat continued to climb, we called it a day. The insects had retreated to somewhere in the Russian olive groves. The birds had quieted down. Time for a midday siesta. I had visions of a cool drink. We walked back to the parking lot.

James was feeling triumphant. "I've got two more females to lay eggs for me," he said. He sounded as though he'd just won the lottery. "Sometimes I come here and spend hours looking and don't find a thing."

I asked if I could take him to lunch.

"I can't. I have to get home. I have thousands of mouths to feed. Literally."

There was not a cloud in the sky, but somehow the atmosphere looked hazy.

"That's smoke. There are forest fires up in Canada, and the smoke is drifting down here."

"All the way down here?" I was surprised.

I shouldn't have been. Several years earlier, smoke from Siberia had made it all the way to Washington State, carried by the prevailing winds.

These fires we were talking about had not remained in Canada. The next day I drove down toward Portland, Oregon. The smoke thickened with each passing mile. I'd been looking forward to my first glimpse of the Columbia River Gorge. It's said to be a fabulous site.

I wouldn't know.

By the time I got there, smoke had made the gorge barely visible. I passed by cliffs and ridgelines with glowing embers. I'd hoped to do some hiking in between interviews, and as a preliminary adventure, I walked one of the short climbs on the side of the river gorge. The trail began at a popular starting point, and I was annoyed by feeling I was in a line of tourists. But by the next day the fires were covering the cliffsides and the mountaintops, and there would be no hiking anywhere in the gorge region for weeks. Several hikers who only a day later climbed the trail I had climbed had to be rescued.

How would the migrating butterflies navigate in all this climate chaos?

Ten

ON THE RAINDANCE RANCH

Habitat loss does not exist in isolation.

Nick Haddad, The Last Butterflies

I was headed down to meet Amelia in person. She lived in Corvallis, Oregon, and she was by then six years old. Her mother, Molly, had volunteered to show me around the Willamette Valley and talk butterflies, one of her favorite subjects. Amelia's dad, who worked for the federal forest service, had phoned earlier. His scheduled activities had been curtailed because of fire and smoke.

That fire was part of a group of fires called the High Cascades Complex. Begun in late July by several lightning strikes, the complex of fires would not be brought under control until the middle of October. People east of the Willamette Valley had been ordered to stay in their homes or to evacuate entirely. The fires had not yet come to Corvallis (they would eventually) but they were raging just to the south.

For the time being, though, on this second-to-last day of August, the way was clear for us to keep to our agreed-upon schedule. First up was a visit to Raindance Ranch, a 250-acre experimental property owned since 1992 by Warren and Laurie Halsey. The Halseys lease out acreage to local farmers, but have also returned some acreage to its pre-homesteading natural state.

Ten thousand years ago, post–ice age floods had filled this entire hundred-mile-long, thirty-mile-wide valley of the Willamette with a 400-foot-deep lake. Floodwaters from the Scablands had rushed down the ancient Columbia, then arrived at a sharp bend, at what's now Portland. There the water took the easy way out, quite literally, and made a left turn into the valley. The deluge filled up the basin and sloshed around as though the valley were a mere bathtub. Carried along in those roiling waters were the soils and silts that had once covered the entire Northwest. When the water settled, the drift settled on the lake bottom, forming deep, rich earth.

Life thrived because of these soils. The ancient people living in this paradise would have enjoyed a good life. There were tall-grass prairies with an abundance of wild game, plenty of wetlands for migrating birds, fields and forests providing fruits and nuts and root crops for the taking. The earliest residents knew how to remove the toxins from acorns in order to make bread, how to manage fires to keep the grasses short and the open areas productive. Archaeologists have recently found a cache of obsidian biface axes that may be as old as 4,000 years.

One thing the land was not good for, however, was traditional European-style farming. The ancient lake had drained, but the valley floor, surrounded by mountain ranges from which flowed multitudinous rivers and streams, remained quite wet. The soils never truly dried. During the winter rains, the valley hosted slews of temporary ponds, vestiges of the 10,000-year-old lake. Some of these dried on the surface over the summer, but down below, the soils were still too wet to grow European-style crops.

Indigenous people lived with these rhythms, moving with the seasons, following the game, gathering food plants at appropriate places during appropriate times. European farmers employed engineering strategies in order to try to control nature. Homesteaders were tied to one particular parcel of earth. They could not migrate seasonally as the weather changed. Instead of living with the land, they had to become the antithesis of beavers. The requisite draining of the valley land began on a small scale. Techniques for large-scale engineering came of age in the early twentieth century. Today the region is filled with thousands of miles of plastic piping, the sine qua non of industrial agriculture.

"All of this is supposed to be under water in the winter," Molly explained as she drove, pointing out fields that looked dreary and worn. The dusty-dry ice age soils whirled like dervishes above the fields, creating mini-tornadoes. "What you're seeing here is an amazing piece of industry. The entire Willamette Valley used to be one giant flood plain, but that was before the river was channelized."

Today, the Willamette River stays where it is told to stay and does what it is told to do (mostly). It is a mere shadow of its pre-colonization self. Molly and Amelia and I drove past mile after mile of hazelnut farms, almost all of them recently installed. Thanks to modern drainage techniques and super-cheap plastic materials, the valley is now becoming a world capital of hazelnuts, which are supposed to have "anti-aging" properties that give you "perfect skin."

"What's with the hazelnut plantations?" I was mystified.

"California water restrictions meant that lots of their nut plantations have dried up," Molly explained. "Now they're coming to Oregon."

We were watching heavy machinery dig up the soil and unroll massive PVC piping systems that would drain excess water. I had seen this kind of thing done in Provence, in southern France, to take advantage of the rich silts left from Rhône River overflows. But those water-engineering projects, begun in Roman times, had required centuries to perfect. This was occurring instantaneously.

One thing was clear. Monarch butterflies would not thrive among the nut plantations. There was no milkweed. Nor would other butterflies. The nut plantations, as with most monocultures, were devoid of the flowering plants—considered "weeds"—that insects require to survive. At least, not unless the property owners adopted David James's "Beauty with Benefits" plan.

We arrived at the ranch house. Only a small part of the property sits on the valley floor. As we drove uphill, we could easily see below us plenty of industrial farm machinery and the requisite dusty Sahara-like whirlwinds. It looked like a mini–Dust Bowl. For acre upon acre, there was no ground cover. Who knew where these precious ice-age soils would eventually end up? With each flurry of wind, the Willamette was diminished.

Perhaps it was just my imagination, but once on the property, the temperature immediately seemed more reasonable. At least, without all those whirling dervish dust devils, it no longer *looked* like I was in the Sahara.

Tall native grasses abounded. Much loved worldwide by at least forty different species of butterflies and moths, tufted hairgrass turned golden as September prepared to roll in. Deliberately replanted years ago, it was thriving. Umber skipper caterpillars love it, as do deer, elk, and a variety of other grazers. Wapato and camas, wild foods, have also been replanted. Many other native grasses had also returned.

"The Halseys are trying to unengineer highly engineered land," Molly explained.

It's taken time, but the Halseys are patient. They have found that after taking out the drainage pipes and allowing the natural hydrology to take over, indigenous plants have begun returning. The seeds are still there in the ground. All that's been missing is water.

The Halseys planted milkweed long ago and had for years been raising and releasing monarchs just because they liked them. Molly and Amelia had come to talk to them about David James's tagging program. They brought along a few tags for good measure. Several monarchs raised by the Halseys had just emerged from their chrysalises that morning and were waiting, restless and fluttering, in their glass jars. We brought the mesh-covered containers out behind the ranch house, where a profusion of flowers, grasses, butterfly bushes, and trees flourished in the sunlight.

Elsewhere, fires were roaring across the land, but here conditions seemed good for a monarch release. If this insect migrated, would she make it through the smoke? we wondered. But we proceeded. Protecting her in the great beyond was not our province.

After placing a tag on the mitten cell as prescribed, Amelia let her butterfly rest on her finger. The little insect sat there for a bit, as though startled by the sunlight. Then she rose into the air and sat on some rafters extending from the house.

I'd expected her to be gone in an instant, but she wasn't in a hurry. Eventually she flew a short distance to some flowers and rested again. She stayed there so long that we left her to it. With plenty of nectar, there was apparently no particular reason to rush.

Monarchs are not the only butterflies to enjoy Raindance Ranch in recent years. The valley floor acreage owned by the Halseys had during the post–Civil War homesteading period been called the "Gospel Swamp." Ultimately, it was drained and converted into marginal cropland that could grow things like rye, a much more tolerant crop than fussy wheat. When the Halseys bought the property, they let the bottomland flood again. A government program provided large machinery that dug a series of five-acre shallow ponds over the sixty-six-acre wetland.

"It was very dramatic. The land was bare at first," Laurie Halsey told me. "But then the natural seed bank just kicked in and plants just started growing. In a few years, the area was bushy and green. In the springtime dogwoods burst out in leaves. The wild roses shine."

The revitalized wetlands drew butterflies, including the Fender's blue butterfly. Endemic to only the Willamette Valley, this tiny little thing lives a life as different from the life of the monarch as it's possible to imagine. The males are a scintillating, shimmering blue and the females a rather innocuous but protective brown. Some of these small blue butterflies are rarely noticed by the average person, but they are part of an amazingly intricate chain of life that provides the foundation on which we mammals depend.

The Fender's blue is a homebody. Monarchs may travel thousands of miles in only one lifetime, but these little things rarely go far. With a tiny wingspan of no more than an inch, they are weak fliers. They emerge from their chrysalises in May and lay eggs—almost solely on Kincaid's lupine, a rare and cantankerous species of wildflower. The caterpillars eat only young lupine leaflets. As soon as the lupine senesces in July, the caterpillars go to sleep, hiding under detritus, where they stay for nine or ten months, surviving not just the heat and desiccation of the late summer and early fall, but the winter cold as well. When spring returns and the lupine starts to grow, the caterpillars do some more eating, pupate, emerge, fly, and mate. The entire cycle begins again.

The Fender's has had a difficult time weathering the recent changes in the valley. Without Kincaid's lupine, it cannot survive, and the lupine cannot survive when the valley's prairie land is dug up and filled with drainage pipes and planted with nut trees and the like. By now, only about 1 percent of the original prairie remains.

The Fender's blue was discovered and named in the early twentieth century by a mailman in the region who chased butterflies: Kenneth Fender. Several years later it was declared extinct. Then along came twelve-year-old Paul Severn, a modern butterfly fanatic if ever there was one, with a butterfly net on the back of his bike. In 1988, he and a buddy had decided to ascend a mountain near their Oregonian homes, just to see what was there. Severn was, by then, a seasoned lepidopterist with a devotion equal to that of Walter Rothschild. Even then the preteen had memorized the names and important details of wing color and life histories of all the butterflies of the North American continent. He regularly read lepidopterist journals and had a complete collection of the butterflies of Oregon.

Or so he thought.

When he and his buddy followed the old logging road to the top of the mountain, they came across a meadow. There Severn saw, to his amazement, a butterfly he had never seen before. Out came the butterfly net. He brought home several specimens. Checking an old butterfly guide, he found the insect was called a Fender's blue. Nothing was said in the guide about it being extinct. He did not report his find.

Fast-forward a year. Someone told Severn, now thirteen, that he needed to go to a lepidopterist conference. He was thrilled. He had had no idea that others suffered from the same obsession he had.

Off he went. After seeing some Fender's blue specimens, he mentioned that he had just collected his own.

"Impossible," they told him. "The insect is extinct. You have made a mistake."

No one believed him. So home he went, and returned the next day with specimens that turned out to be exactly what he said they were. The hunt was on. The following summer scientists located remnant populations. The insect was then listed as endangered.

Now, several decades later, the Fender's blue is alive and thriving on the Halseys' property, and in many other locations throughout the Willamette Valley as well. In the mid-1990s, the butterfly was thought to number about 1,500 individuals. Today it numbers roughly 28,000. These population figures are rising year by year.

The story of how this seemingly innocuous little butterfly—one of a group of butterflies commonly called "small blues"—went from being categorized as "extinct" to becoming revitalized is a twenty-five-year-long saga that encapsulates how much more we know now about butterflies than Charles Darwin and Walter Rothschild and Herman Strecker and even Miriam Rothschild knew. The insects' complex connections with the living world would have delighted them, and they would have understood quite quickly that it is not enough to merely set aside land for them.

To successfully preserve a butterfly, you need to discover the insect's entire life history—not just what it eats, but where it hangs out and who its friends are. That task can be time-consuming and complicated.

Long ago, to preserve a different butterfly, the Nature Conservancy bought a bog along Washington State's Yakima River. The area was fenced off from cattle grazing. Butterfly protected. Job well done.

Or so it was thought. No one then realized that that particular butterfly species relied on a particular violet species, and that that violet relied on grazing to keep the grasses short. Once the cattle were removed, invasive grasses, shrubs, and tall trees took over. The violets failed to grow. The butterflies disappeared.

So to protect the Fender's blue, scientists realized they needed to know the insect's basic biology. They found that a whole system, like the one described above, had to be intact for the Fender's blue to thrive. Yes, the butterflies needed the lupine. But the lupines needed fire. Once common on prairies, fires are usually ignited by lightning strikes during summer dry periods. Indigenous people burned the Willamette frequently, to keep areas open to draw in the game.

The butterfly also needed ants. Something like a quarter of all butterfly species enjoy special relationships with ants. Some are obligated to be associated with ants, while others merely do better when certain species of ants are around to help out.

The Fender's blue caterpillar has special organs that secrete a sweet liquid craved by certain ants. Once the ants find a caterpillar, they are like kids in a candy store. They can have all the candy they like, as long as they can keep other ants and predators like wasps away from "their" caterpillar. Which is exactly what the ants do and the caterpillars need.

The ants are bodyguards. They protect the caterpillar from other species that might do it harm, in order to keep the candy store open. Fender's blues do not *have* to have these ants present and protective, but they survive in much greater numbers when the ants keep the bad guys away.

So to survive, the Fender's blue needs not just land—but land with the right kind of lupine. That land also needs to burn periodically. Given the lengthy list of requirements, experts assumed that large swaths of land would need to be purchased. That was unlikely, given the price of Willamette farm land. Then along came Cheryl Schultz, a newly minted scientist looking for a project that would make a difference not just in population numbers for a specific species, but in overall approach to wildlife protection.

Schultz and her colleague Elizabeth Crone figured out that it takes a village to save a butterfly. They worked the numbers and realized that the Fender's blue did not require extensive expanses of land—but merely *small parcels* of land placed a few kilometers distant from each other. Then the butterflies could use this chain of refuges as stepping stones. Several acres here and there would allow the weak-flying insects to spread. Populations blossomed on several publicly owned properties—Molly and Amelia and I went to see a few of these places—and conservationists found private landowners willing to farm for the butterflies. One local vintner now markets a "Fender's Blue Red."

• • •

So it takes a village—five-year-old girls and their parents and caring landowners and volunteer scientists and marketing-minded vintners and painstaking researchers—to save a butterfly. The standard for how to save a butterfly was set in 1979 in the United Kingdom and Europe. These scientists, too, were working on how to protect another vulnerable small blue butterfly, a cousin to the Fender's blue.

The butterfly in question, nicknamed the "large blue," had peculiar tastes, cannibalistic tendencies, and a mysterious lifestyle. (The word "large" was meant comparatively. The Fender's blue has a wingspread of about one inch, while the large blue has a wingspread of about one and a half to two inches.)

Once spread across Northern Europe and Asia, the large blue was never common in the United Kingdom, but it was treasured. It was the subject of ongoing, in-depth discussions in *The Times* of London.

It could not be hand reared. No one knew why.

For butterfly collectors, the large blue was simply irresistible. Wings of both the male and female were a glistening royal blue. They flashed with dazzling brightness, like neon signs. At the wings' edges ran a thin border of black. It would have seemed funereal, were it not edged again by an even thinner elegant border of pure white. The forewings hosted an arc of several dark spots, which some people described as "teardrops," with a half moon inside that arc.

Imagine the scene: After a dark and dreary Northern European winter, Victorians flock to the countryside to enjoy lollygagging and picnicking. Blankets are spread. Food is consumed. Wine and beer are plentiful. The bright sun is almost hot. People stretch out luxuriously, enjoying the greenery. For recreation, there are a few butterfly nets. The large blue's flight time of just a bit more than a week comes at midsummer, during summer's longest days.

But by the 1920s, the large blue had nearly vanished from Britain. Collectors were blamed—incorrectly, it later turned out. The solution seemed to be to fence off areas where the butterfly flew, forbidding entry to humans and to grazers like cattle and horses. Give the butterflies their space. That oughta do it. Sounded like a good plan.

It wasn't.

The problem worsened. In 1979, the butterfly was declared extinct in Britain. Elsewhere across northern Europe, populations also declined. Oddly, though, the species hung around where traditional herding and grazing practices continued.

Researchers set out to learn why. If understanding the complexity of the Fender's blue's survival was difficult, the large blue required a tightly knit labyrinth of connections, like the twists and turns of popcorn stitches on woolen sweaters. This wasn't just about one butterfly and one plant. This was about a whole system.

It took thirty-five years to find all the puzzle pieces and assemble the entire picture. Large blues are cursedly picky—"neurotic aristo-crats," according to the famed British butterfly fiend Matthew Oates. When large blues' caterpillars hatch from eggs in early summer, they eat through the flower heads of wild thyme, feasting on their energy-rich seeds. They also fight to the death. When two caterpillars meet, both adopt a take-no-prisoners strategy. The winner cannibalizes the other.

The caterpillars eat the flower heads only up to a certain age. After that, they drop off the plant. They wait on the ground, like hitchhikers standing by the roadside.

Along come red ants, the kind that normally predate on vulnerable things like young caterpillars. But in the case of the large blue, the ants instead crawl all over the caterpillar. They hoist him up and carry him home, like a wounded hero.

In the nest, the caterpillar will, using various methods, essentially disguise its true nature and try to become one with the ants.

Zen caterpillar.

Then things become interesting. Nurse ants treat the caterpillar like royalty. In turn, the intruder behaves quite like a queen ant and settles down for a long winter's nap.

Waking, it feasts on the ants' offspring.

Not so Zen anymore.

Nine months later, the well-fed and immensely coddled caterpillar pupates in the ants' nest and emerges as a butterfly. Guard ants parade the butterfly out of the nest, as though partaking in a royal procession, and wave it on its way.

The caterpillars survive in some ant nests but not in others. Why? Researcher Jeremy Thomas looked at all the red ants living in the areas around which large blues flourished. He found that there were in fact five different species of red ants in the area (who knew?), all of which, to the casual eye, looked alike.

But the large blue's caterpillars were attuned to the lifestyle of only one particular species. If they were carried into the nest of that species, they thrived. If carried into the nest of the wrong species, they were toast.

Why did the ants of this one species give the caterpillar a hero's welcome? Researchers found two reasons. First—and this is *wild*—the large blue's caterpillars exuded a compound that mimicked the compound used by the ant species as an identifier. These ants recognized other ants of the same species by detecting this specific compound. So when they encountered the compound, they responded as though the caterpillar were a wounded colleague.

Second—and this is even *more* wild—the caterpillar mimicked the ants' *sounds*. These sounds was apparently a siren's song calling to the ants. The caterpillar wasn't just passively waiting at the bus stop for transportation. It was *summoning* the transportation.

Some researchers believe that the reason the caterpillar is treated so well by the ants is that the caterpillar makes the sounds made by *queen* ants. If there is a real queen in the nest, this does not go over too well. But if the ants don't have a queen, some researchers believe, they accept the caterpillar as royalty.

The tricks the caterpillar plays on the ants are deadly serious. If the pretending caterpillar is successful, it will thrive. But if its deception is discovered, the caterpillar will be eaten. This turns out to happen rather frequently. Only the best mimics survive. Darwin would have loved this story.

Once the system was understood, the next step was to figure out what had broken down. Scientists studied the ants that unknowingly partnered with the butterflies. These ants turned out to be fussy Goldi-locks types. They didn't like things too hot. They didn't like things too cold. Temperatures had to be just right or other, less temperamental ants would flourish instead. The ants did not do well with too much rain. Nor did they do well with too little rain.

The requisite wild thyme turned out to be another glitch. That plant needed *its* own support system. The key here was the array of grass spe-cies. Suburban lawns? Forget about it. A variety of plants were needed.

But the grasses could not grow too tall. Enter the rabbits. They grazed the grasses down to levels good for the butterflies. *But* a virus called myxomatosis had killed off most rabbits.

Hence the grasses grew tall. The thyme disappeared. The ants failed in their duty. The large blues in Britain died off.

Here scientists were stymied. The rabbits were gone, and few peo-ple wanted them back. What to do? They realized it wasn't the rabbits themselves that were needed. It was grazing that was key. Why not graze cattle and horses? They would eat more grass with every bite, and would be much easier to control, in part because they wouldn't breed like, well, rabbits.

Experiments showed that managing grass heights in this way would be effective, but even this was complicated. The grazing had to be managed. You couldn't just leave the horses and cattle in the fields for months on end. Livestock needed to be put on the fields at precisely the right times, and taken off at precisely the right times. It was a House of Cards system.

Strains of large blues (some people think of these strains as subspe-cies) were taken from locations outside Britain and reestablished on the island. At last all the various strands had been sorted out, and this time the effort was successful.

To get a sense of this butterfly's power over the British psyche, Oates tells the following story: When a refuge for large blues called Collard Hill was finally open to visitors, an older fellow about to have a double hip replacement arrived. The butterflies lived on quite a steep hillside. To see them, he had to climb down and then up a difficult slope. He was not deterred. He had flown fifty missions over Nazi Germany. The spirit was still in him. When he got down the slope to where the butterflies lived, "a Large Blue settled to bask right beside him, and in that moment he fulfilled his lifelong ambition to see every species of British butterfly."

Today, this species seems relatively secure, although conservationists may never be able to relax their vigilance. The human dimension remains a problem. Only recently a smuggler was arrested with a number of large blue specimens in his possession. He had planned to make a killing by selling them on the international underground Lepidoptera market. Caught out by a citizen conservationist, he received a visit from the police, who busted him in his house.

At the trial, Neil Hulme, project officer for the nonprofit Butterfly Conservation, said that in the twenty-first century butterfly rustling is no longer widespread, but "the people involved are quite determined."

The addiction is alive and well and flourishes in the human brain.

Some say that the large blue is Britain's most successful conservation effort. This may well be true. They also like to say that this is the first time anywhere in the world that a butterfly has been successfully reestablished in a location where it had been eradicated. The save might have occurred just in the nick of time.

Other members of this delicate group of butterflies have already disappeared. The best-known example is the Xerces blue. First described in 1852, it was last seen in the 1940s. Its sole place of abode had been the sand dunes along the Pacific coast of San Francisco. The plant it required had been exterminated by streetcars, which carried people from their city center day jobs to their suburban homes—in this case, to the city's newly emerging Sunset District (ironically, the district where Amelia's butterfly would turn up decades later, taking advantage of a roof deck flower garden).

The same might have happened to a small blue butterfly known as the Karner blue, had it not been for a spectacular rescue story in New York State, near the state capital of Albany.

Eleven

A SENSE OF MYSTICAL WONDER

[T]he highest enjoyment of timelessness . . . is when I stand among
rare butterflies and their food plants. This is ecstasy, and behind the
ecstasy is something else, which is hard to explain.

Vladimir Nabokov, Speak, Memory

This story begins a century earlier, deep in the Russian countryside.
Renowned author and amateur lepidopterist Vladimir Nabokov,
born in 1899 just at the end of the Victorian era, had a reverence for
butterflies that exceeded perhaps even that of Walter Rothschild. His
passion began almost in infancy, when Nabokov, following literally in
his aristocratic father's footsteps, started to learn to identify myriads of
butterfly species. By age ten, he was absorbing international scientific
journals.

He had also formulated his central life goal: to name a new butterfly
species. Around that time, he sent a letter to a journal announcing the
discovery of a "new" species, only to be shot down by the journal as a
mere "schoolboy." Sadly, the species had already been described.

Nabokov loved the butterflies on his family's estates. Serfs netted
them for him. Just as Miriam Rothschild's father, Charles, once stopped
a train in order to have his servants catch a coveted butterfly he had
seen out the train window, when Nabokov was seven he caught sight of
a butterfly and had a servant catch it. At that age, his first thought in the

morning was wondering what butterflies he would see that day. About one butterfly he saw as a child, he wrote: "My desire for it was one of the most intense I have ever experienced."

This craving was hereditary, he explained in his exquisite *Speak, Memory.* "There was a certain spot in the forest, a footbridge across a brown brook, where my father would piously pause to recall the rare butterfly that, on the seventeenth of August, 1883, his German tutor had netted for him." His memoir includes a map of the estate showing the location of a prized catch. The father's enthusiasm was passed on to his son, and their shared passion created an irresistible bond. When the father was sent to prison for defying the czar, the two sent letters back and forth discussing butterflies. By letter, Vladimir learned about a butterfly his father had seen in the prison yard.

When the revolution came, his aristocratic family escaped and ended up penniless in Germany. After Hitler took over, Nabokov himself wound up in Boston, teaching at Wellesley College. Eventually he taught Russian literature at Cornell. Following the smash success of the scandalous *Lolita*, Nabokov became the world's most famous lepidopterist. Journalists loved to write about his butterfly fascination, usually portrayed as revealing his esoteric artistic temperament. Quite often the magazine photos included stills of him with his net.

While at Wellesley, Nabokov moonlighted at Harvard's Museum of Comparative Zoology. He received an appointment there. Fascinated by the hidden diversity of the small blues, he devoted himself zealously to his work, carefully dissecting specimens so as to study their genitalia. (This wasn't necessarily salacious: lepidopterists commonly study genitalia in order to determine, among other things, the sex of butterflies.)

One reason why he was spellbound by Lepidoptera was his special relationship with color. For Nabokov, color was omnipresent. Letters of the alphabet came in specific colors. "In the green group," he wrote, "there are alder-leaf f, the unripe apple of p, and pistachio t. Dull green, combined somehow with violet, is the best I can do for w." He seems to have inherited this from his mother, also a synesthete.

It is no wonder, then, that for Nabokov the brilliant flutter of butterfly wings in the summer sunlight evoked a sense of mystical wonder. The language of butterflies was a language in which he was innately proficient.

Nabokov loved the small blue butterflies with their highly refined lifestyles. In the American Northeast, he was particularly interested in one little blue butterfly, but never seemed to be in the right place at the right time to see it. Then, driving one summer between Cornell and Boston, he found a field of lupine that was replete with his yearned-for species.

The butterfly, he determined, was an unnamed species. He called it the Karner blue, after the little New York village railroad stop where he'd found it. Its formal Latin name has ever since included "Nabokov" at the end, indicating that he was the describer. His life's goal achieved, he would call himself "godfather to an insect."

The Karner blue was once common. Observers had written about the "clouds of blue" that lifted off when disturbed. But even as Nabokov discovered it in the 1940s, numbers were dwindling. By the 1970s, those clouds no longer existed. Concerns were raised, but not much happened regarding its conservation in New York until a land developer proposed building a shopping mall in the area. Advocates spoke up.

An epic battle raged. Ultimately, all parties compromised. The shopping mall was built, but several hundred acres of land were also set aside for habitat revitalization. A legal order called for the restoration not of the Karner blue specifically, but of the entire ecosystem where the butterfly thrived.

I had heard that the revived ecosystem was spectacular. I decided to go for a walk.

No sooner had I parked my car at the Albany Pine Bush Preserve than I found myself face to face with a monarch on a milkweed plant. Flashing around in the late-summer sunlight, it gleamed and glistened, shimmered and twinkled. The visitor center is a rehabilitated bank building. Its parking lot, once an expansive stretch of nothing but pavement, has also been reborn. Lupines, the Karner's obligate plant, now grow where pavement used to be.

Raised beds containing native plants, including lots of milkweeds, draw in myriads of birds, insects, and little mammals. Each of these native plants alone might be considered a "weed" in suburbia, but growing alongside all these other native plants they create a glorious lushness filled with color and sound and with a natural vibrancy that's long gone from many of our landscapes.

"Today's a great day for the Albany Pine Bush," Neil Gifford, the preserve's conservation director, told me as we shook hands. "It's all about the ecosystem here."

I had shown up on the day that Gifford and others were announcing a triumph: the local Karner blue population had risen from a dire 500 or so specimens in 2007 to around 15,000 in 2016. This wasn't an anomaly. The healthy population figures had been consistent for the past several years.

Before Gifford and I sat down to talk, I walked over the landscape for several hours. Miles of dirt roads and foot trails lace the land. The Preserve, a joint project managed by several municipalities, was initially only a few hundred acres. It's now more than 3,300 acres in size. Gifford wants 5,000.

As I walked, around every bend, on top of every hill, new sights and sounds caught my attention. One shallow pond, cacophonous with croaking, reminded me of places I had seen in Africa. Birds were everywhere. Butterflies filled the air. More than 20 rare species live here. Additionally, there are more than 90 bird species, fishers, several kinds of turtles, lots of snakes, at least 11 species of trees, honeysuckle and bracken and grasses and sedges galore. There are so many different types of wildflowers that, except during the winter months, there is always something in flower. Gifford later told me there were at least 76 wildflower species designated as in need of protection thriving in the preserve.

Plenty of people enjoy the area, too. These lands are not just for wildlife. The dirt roads and paths that I walked are open to all kinds of recreation: walking, of course, and bicycling and horse trekking, cross-country skiing, and even, at times, hunting.

All this, I marveled, alongside one of the busiest travel corridors in the country—Interstate 90. I could hear the rumble of semis and the constant clatter of horns and sirens and stop-and-go cars. And yet, I still felt I was walking in an area filled with natural vitality.

The key to the Albany preserve's species wealth is fire. Because of lightning strikes, the area used to burn frequently and naturally, over thousands of years. Paleoindians hunted here 10,000 years ago, and evidence from pollen samples proves that people were burning before Europeans arrived. Paleoindians were likely actively managing the area by the end of the last ice age.

Gifford explained: "Species here are completely fire-dependent. They are not only adapted to deal with recurring fire, but in many cases they have adaptations that *require fire.*"

Pitch pine and jack pine cones are closed until fire melts the resin and releases the seed. The fire also preps the ground with ash, nourishing it for the opened seeds.

"Most people didn't expect the butterfly to do this well," he said.

This project is very much his baby. He's devoted his entire working career to stewarding this land, much the way a farmer would devote his lifetime to the improvement of his farm.

"Amphibians and snakes have also exploded as a result of our management. Birdsfoot violet, a gorgeous little violet with a big beautiful flower, now grows here. I did not realize that this species of plant was here. It must have been in the soil's 'seed bank,' just waiting. When we started burning it was amazing to see the response from the little violet, which supports the regal fritillary," a somewhat rare butterfly with orange wings.

Also thriving here now is New Jersey tea, so named because colonists used it for tea during the Revolutionary War. It throws its seeds into the soil, but "only after a fire will those seeds actually germinate," Gifford said. The plant prefers sandy soil, grows to only about two feet in height, and produces lush clusters of flowers that draw a variety of birds and insects, including several species of butterflies.

It took quite a while for scientists to realize that fire—and only fire—would bring the plant community back. For well over ten years, the land was preserved but fire was not used. The Karner failed to flourish. Incorporating fire into a system surrounded by people's homes was daunting.

At 3,300 acres, the preserve is huge for an urban park, but that acreage is not contiguous. It is broken up into sometimes larger and sometimes smaller sections. In between are housing developments, shopping areas, and busy highways. Gifford and others had to find ways to burn these parcels of land while not disturbing people.

Maintenance fires must be done routinely: "This is relatively frequent, but low intensity. We were the first to refine the ability to use prescribed fire in a fragmented urban landscape like this, where there is so much edge and development. There is no place to put the smoke. We have no leeway. We can't dump smoke on the roadways or on other people's property. The fire can never get away from us. We have no flexibility."

Because of this, prescribed fires are never allowed to burn on more than fifty acres at a time. "We have to be careful that we are not biting off more than we can chew. We have to mop up by sunset."

Where, I asked him, was the money coming from to make this happen?

He pointed in a specific direction.

"There," he said. "Mount Trashmore."

It was, in fact, a mountain of trash that he pointed to, sitting just adjacent to the high-quality preserve land. I had noticed it as I was walking, but had not understood exactly what it was.

The City of Albany has been allowing other municipalities to dump their trash on this growing heap for years. For this privilege, municipalities pay Albany. The court order requires that some of that money be given to the preserve.

The sand dunes that provide the foundation for the Albany Pine Bush Preserve are a gift to us from the Pleistocene ice ages. So says the area geologist Robert Titus: "It's a human notion, a gift . . . but so many things that we humans value in this region date back to the effects of this ice age—the beautiful Catskill landscape, the art, the literature, all date back to this." At the end of the Pleistocene, when the ice started melting, a lake of water formed at the base of the glacier. Glacial Lake Albany stretched all the way south to a town now called Beacon, just north of New York City.

The lake was fed by tributary rivers and streams, including an ancient version of the Mohawk River. A delta, now long gone, formed at the mouth of this river. The lake's recession after The Great Melt exposed the delta to whirling, frigid winds. Those winds picked up the sand and lighter materials of the delta and blew them east, creating walking sand dunes like those of the Sahara Desert.

"It's quite something to imagine," wrote Titus in *The Hudson Valley in the Ice Age.* "For a substantial period of time, this part of Albany was a cold-climate desert. Large, treeless dunes migrated across the countryside, driven by the wind. Toss in a few camels, and it certainly gives you a different impression of Albany." Of course, as far as we know, there were no camels in the region at that time, although there were plenty of camels then ranging elsewhere across North America.

Fires burned here often, set off sometimes by lightning strikes. The first people, like their associates in the Willamette Valley, used fire to maintain the land and to help them hunt. By setting fires, they were able to keep the land clear by keeping the forests from returning. European-based land ownership systems put a stop to those fires, as in the Willamette Valley.

Like the sand dunes of Albany, now mostly gone except for those in the preserve, the entire group of little blue butterfly species is also a gift to us from the ice ages. It was Vladimir Nabokov himself who first proposed this. In a paper published in 1945, Nabokov suggested that the butterflies flew from west to east, riding on the prevailing winds in the Northern Hemisphere, in five successive waves, beginning almost 11 million years ago and ending about 1 million years ago. The pattern he suggested later turned out to correlate with global climate change patterns.

In 2011, using DNA, an international team of ten scientists confirmed that Nabokov had been spot on. Shifting tectonics and changing climates encouraged the spread of these species, which then adapted to whatever was available for them.

The small blue butterflies live unique, insular lives, locally focused, that depend upon an intricate set of highly specialized connections. Once we understand them, it is possible to conserve them, if we care enough to set aside space on the planet and invest money.

But what about a species like the monarch—a species that roams thousands of miles, that migrates and requires healthy habitat all the way from the Canadian grasslands to the Mexican mountains?

Part III

FUTURE

Twelve

THE SOCIAL BUTTERFLY

Tomorrow may rain, so
I'll follow the sun.

The Beatles

Kingston Leong was a disappointed man.

It was only days before Thanksgiving 2017, about eight months after we had first met under the overwintering trees of the central California coast. Leong looked pensively at the grove of trees he had planted so long ago. We were once again at the Morro Bay golf course. The previous year he had counted 17,000 monarchs, down from 24,000 the year before.

The time had come for another census. Hope springs eternal: certainly monarch numbers would have risen. The fires of the Pacific Northwest had not reached here. Nor had temperatures along this coast been extreme. Instead, the rains of the previous winter had nurtured a luxuriant carpet of wildflowers, including the all-essential milkweed. He expected to see many butterflies.

He didn't.

Together we looked in all the places where the butterflies typically congregated. On branches where last winter there had been clusters hundreds thick, now only a few were visible.

"There," he said, "over there, where it looks like a branch of dead leaves."

We walked over to take a closer look. It *was* a branch of dead leaves.

Then we began to see them dancing in the sunlight. Rather than clustering, many were flying or spreading their wings to take in the sun's warmth. It was early in the morning, when the insects were usually huddled against the cold.

But today was a disaster: a pitch-perfect California day. Sunny, mid-seventies, no wind. All California seemed to party in celebration of the unseasonably glorious weather—all, that is, except for monarch monitors trying to count insect numbers.

Rather than huddling, the monarchs were out and about, flitting here and there in total disregard of their appointed task: nestling together to survive miserable winter weather. We saw some landing on a small branch where others were resting, then saw all of them rise together and take off. Many were stopping to fuel up at the grove's vast carpets of ground-hugging succulents, ice plants, which offered expanses of both purple and white flowers. Despite the fact that it was November and we had arrived quite early, the butterflies were carousing as though it were summer.

Butterflies are little solar panels, powered by the sun. It looked almost as if on such a day, they *had* to fly. Glorious to behold, this was calamitous for the insects. The problem with flying behavior is that it uses up energy. On their migration to the coast, the butterflies would have consumed as much nectar as possible in order to store up fat for the winter. The migration expert Hugh Dingle suggests that monarchs can store up to 125 percent of their body weight as fat, but this kind of seemingly frivolous flight behavior would use up much of those essential reserves, and there were few flowering plants open for the business of refueling at this time of year.

Something was wrong. Leong decided to return after the New Year, when the weather was likely to be suitably dismal. He continued to hold out hope that the numbers would be high, but when he returned in early January he found only 13,000.

Larry Gall with one of Yale's butterfly boxes.

A monarch spreads her wings to rest in the sun.

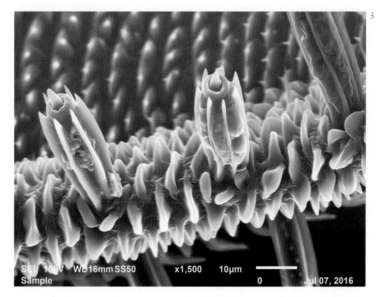

The Atala butterfly, once thought extinct, has made a remarkable comeback. Taken with a scanning electron microscope, this photograph shows the tip of the butterfly's proboscis. The vicious-looking spikes actually contain nerves that "taste" the surface of a possible food almost the way our own taste buds taste food.

The proboscis tip of a hackberry butterfly, with its taste bud–like spikes, taken with a scanning electron microscope. When the butterfly lays its proboscis down on rotting fruit, the food moves into the darker spaces via capillary action, then moves farther into the food canal.

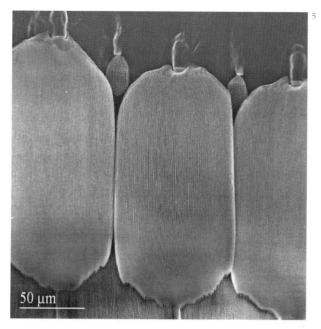

In this photo of individual scales of a blue morpho, characteristic ridges are formed by the peaks of the "Christmas tree" structures.

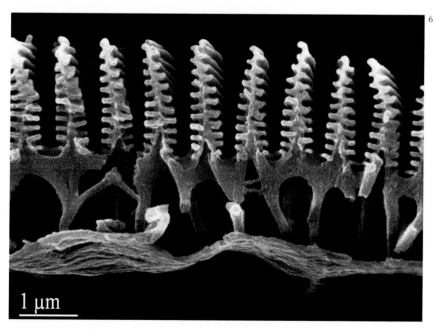

1 μm

A cross section of a morpho scale shows the Christmas tree structure that created the astonishing beauty that mesmerized Maria Sibylla Merian.

Paleontologist
Gwen Antell.

Entomologist
Matthew Lehnert.

Fossil collector Jim Barkley helps young paleontologist Hans Snortheim prepare a just-collected fossil for exhibition.

Jim Barkley's research site.

Ecologist Anurag Agrawal explains one of the finer points of the milkweed-monarch partnership.

Normally female monarchs lay one egg on the bottom of a milkweed leaf. In this unusual case, a female has laid three eggs quite near one another on a milkweed bloom. Citizen scientist and photographer Carol Komassa suggests that this might be an example of "egg dumping," which occurs when a tired female at the end of her life wants to deposit as many eggs as possible before dying.

A slightly tattered blue morpho rests on the author's jacket.

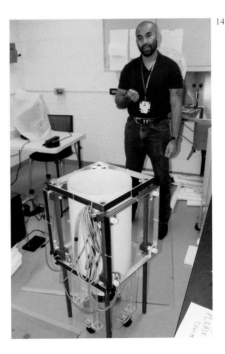

Monarch researcher Patrick Guerra uses an open barrel to study monarch flight orientation.

Chip Taylor and Andrew Gourd
of the Seneca-Cayuga people
confer over the collection of
native wildflower seeds.

Monarchs cluster
while tourists watch.

Monarchs cluster on
the ground in
the Mexican sunlight.

Monarch souvenirs for sale at the entrance to the walking path in the El Rosario Monarch Butterfly Biosphere Reserve in Michoacán, Mexico.

Walking up the trail in the Mexican mountains to see the phenomenal site where millions upon millions of monarchs spend their winter huddled together at 11,000 feet in altitude.

20

Nipam Patel, director of Woods Hole's Marine Biological Laboratory, displays one of his boxes of blue morpho butterflies.

21

The newly emerged monarch caterpillar eats its shell as a quick first meal.

22

A tiny translucent caterpillar, barely visible, emerges from its egg.

A particularly spectacular species of blue morpho, *morpho cypris*.

A drop of liquid on the tip of the proboscis, in the process of being absorbed.

Left: Monarch biologist David James explains how to attach one of his monarch tags to a captured butterfly. *Top right:* David James demonstrates how a monarch wears a tag. *Lower right:* David James gently spreads a captive monarch butterfly's wings.

An artist's vision of *Prodryas persephone* enjoying life in Florissant more than thirty million years ago.

The tiny Karner blue butterfly, beloved by famed author Vladimir Nabokov, has been brought back from the brink of extinction at the Albany Pine Bush Preserve in New York State.

A field of wild blue lupine, required by the Karner blue butterfly, flourishes at the Albany Pine Bush Preserve due to a careful regimen of repeated burnings.

Mai Reitmeyer, a research
librarian at the American
Museum of Natural History,
and the author peruse an
original copy of Maria Sibylla
Merian's hand-painted tome on
the insects of Suriname.

Monarch butterflies cluster
for the winter on a tree
branch at Pismo Beach.

A branch of an orange tree, containing both blooms and mature fruit, with a Roth-schildia moth in its caterpillar stage, cocoon stage, and flying stage. "So if someone wanted to go to the trouble of collecting these caterpillars, they could produce good silk and make a large profit." Plate 52 of Merian's work on the insects of Suriname.

The many different stages of *Arsenura armida*, both male and female, shown on the "palisade tree," a tree that according to Maria Sibylla Merian, was used by local people to build houses. "The lower, smaller one is the male," Merian wrote. "The larger upper one is the female." Plate 11 of her work on the insects of Suriname.

A high-percentage decline of 11,000 butterflies at this one small site over the past two years worried him. We discussed possible causes. After an immensely promising spring with plentiful wildflowers and milkweed, the weather had turned quite hot. Rains were sparse. The spring foliage that had looked so beautiful had turned tinder-dry. The wildfires of August burned all the way into September and then into October, spreading south. California experienced some of the state's worst fires in recorded history.

Although there were no fires near the golf course, it seemed likely that many of the wildflower fields upon which the monarchs depended for fuel throughout their migration had burned. The fires may also have killed large numbers of migrators directly. Or the smoke, so overwhelming that even people living in usually foggy San Francisco had to wear face masks, could have interfered with the insects' complex and delicate navigation systems. (There is some disagreement among scientists over this question.)

Or, it could be that the numbers were healthy, but the insects had gathered elsewhere, somewhere that no one knew about; or . . . The list of possibilities we discussed seemed endless.

Over the next several days I returned to the other sites I'd visited the previous February and found the same thing. At Pismo Beach I joined a training session for volunteers who were learning to count clustered monarchs. We met quite early in the morning. The air was chilly. Few insects were flying.

An experienced team had already finished their estimate: 12,382 individuals.

"I'm not going to lie. It's low," responded monarch biologist Jessica Griffiths. Indeed it was—much lower than in years past. In earlier years of this century, numbers had ranged in the tens of thousands. In the decade prior to that, numbers had been above 100,000. In the winter of 1991–1992, an estimated 230,000 butterflies had gathered here.

Causes for the decline are multitudinous. Some are mysterious: How much did the climate chaos I had continuously witnessed interfere with the monarch life cycle? How did the smoke from the constant fires affect them? How much did the fires themselves, which would have destroyed nectar-filled flowers, harm the migrating butterflies?

On the other hand, some causes are simple: when I visited Pismo Beach for the Thanksgiving count, it turned out that one of the butterflies' favorite eucalyptus trees had fallen in late spring. Eucalyptus trees have a lifespan of about a century, so this wasn't unusual, but the winter's heavy rains had made things worse. Most of the eucalyptus root system grows in the top twelve inches of soil, which is easily penetrated by heavy rains.

The result is destabilization. That tree had fallen against another tree, knocking it over also. The domino-like demise of the two trees had left a large hole in the grove, which changed conditions, allowing more wind to flow through the grove. A forest is a dynamic place. Nothing stays the same. In times past, the monarchs would most likely have simply moved a bit sideways, to another nearby seaside habitat that had the right qualifications. However, with increased construction along the seaside, those alternative sites are disappearing.

A mother and teenage daughter had come to the training session with news of a small overwintering site that no one knew about. Griffiths and her colleagues visited later and found a small number of the insects, but could not tell whether it was a refuge that monarchs had been using all along, or whether the monarchs were there because the Pismo Beach site was no longer satisfactory.

After that, I caught up with David James. It was a few months after I'd visited him in Washington State. He was spending the Thanksgiving vacation week with his wife and children traveling up the coast looking for butterflies that had been tagged by his volunteers. We visited a variety of overwintering sites. Some were fairly large. Some were small. Some were on public property. Many were on private property. By now, well over 400 sites of varying sizes have been pinpointed, spread out all along the California coast from as far south as San Diego to far north of San Francisco. More are discovered every year. Meanwhile, other sites blink out.

Mia Monroe, a volunteer who has been counting monarch butterflies for several decades, mentioned her own personal theory: once upon a time the whole of the coastline might have been a migration destination. Butterflies could easily have moved from one grove to another as conditions changed. Over time, however, as people colonized the coastline, Monroe suggested, that one long and continuous site had been broken up by development.

"This is a rapidly developing world of understanding," she explained, cautioning that her idea was only one among many. "I tend to look at monarchs as, instead of being grove-centric, as being region-centric." If they fly to one grove and conditions are not correct, she suggested, they seem to find somewhere new.

"They're insects," she said, "and therefore extremely temperature sensitive." They go where conditions are most suitable at any given moment.

I was struck by her comment. It's easy to forget that insects have no way of internally stabilizing their own temperatures. When we mammals are cold, we can thermoregulate ourselves so many different ways. We shiver when we are cold, some of our blood vessels may constrict, or maybe we move around, increasing heartbeat and blood flow until we warm up. Insects have no such ability.

Unable to survive cold weather, butterflies have devised other strategies. The little blue butterflies that arrived in the New World during the ice ages evolved such clever options as being taken care of as caterpillars by ants in warm underground lairs during cold northern winters.

Other insects opted for migration: when the weather turns cold, many species head south. Research suggests that monarchs evolved at least a million years ago in northern Mexico and the American Southwest. This was a time when northern ice sheets ebbed and flowed over North America, and the climate, as it is now, would have been unpredictable. The monarchs' solution—each year spreading generation by generation north as the milkweed bloomed, and then flying south to safety in one fell swoop of a massive migration when the weather turned—makes good sense.

After all, following the sun is what many humans would do, were we free to do so.

Our discussion made me wonder once again: What guides the migrating monarch? What *inspires* an insect weighing only a few tenths of a gram to undertake such a journey?

And how the heck do they know where they're going?

During the last half of the nineteenth century, when Victorian-era butterfly mania was at its height, observers reported that southbound monarchs migrating in the American Northeast flew in streams that extended for miles and obscured the sun, "blurring day into night," as William Leach sums up an 1868 account by Charles Valentine Riley, a biologist who was among the first to suggest that the insects might be flying long distances. "Millions of them were seen passing for hours, even in Boston," writes Bernd Heinrich in *The Homing Instinct*. Contemporary observers in 1885 and 1896 described monarch numbers as "almost past belief" and wrote that "the heavens became almost black with swarms of huge red-winged butterflies."

Although the fall flight of the monarchs was well known, northerners had no idea where the insects were going. Many people knew that monarchs west of the Rockies migrated to the coast, but no one imagined that the ultimate destination of the eastern and central flyway monarchs was a small region in the Mexican mountains an hour or so west of Mexico City. The idea would have seemed absurd.

The mystery remained unsolved until the Canadian biologist Fred Urquhart, a lover of monarchs since childhood, decided to find out. It took him a lifetime. In the mid-twentieth century he began the first monarch monitoring program. Following World War II, he and his wife, Norah, began developing a citizen science program that spanned the North American continent. Everything was done by hand. In this precomputer era, people who found tagged monarchs reported their locations by snail mail. Then Urquhart marked those finds on a huge wall map. He drew black lines from the point where the individual butterflies had been tagged to where they had been recovered.

Initially, the lines converged in Texas, but no overwintering location was found there. Urquhart concluded that the insects were heading south of the Texas-Mexico border. Few people believed him. Then, a pair of Urquhart's volunteers in Mexico, an amateur naturalist from the U.S. and his Mexican wife, followed the trail of monarch sightings into the Sierra Madre, asked the locals, and thus came upon the mountainous overwintering site. They telephoned Urquhart. "We have found them—millions of monarchs!" The scientist and his team were ecstatic.

But finding a monarch congregation did not *prove* that the butterflies on the Mexican mountaintops were the *same* butterflies that had streamed south from Canada and elsewhere. In fact, the idea seems crazy. How could they fly all the way from Canada to Mexico? And how in the world would they locate these exotic mountain refuges, having never been there before? Buying into the concept required a huge leap of faith.

Sometimes the best-laid plans go even better than planned. Urquhart, by then elderly, decided to make a pilgrimage to see for himself. When he arrived on the mountaintop, a steep climb from 11,000 to 12,000 feet, he sat down to rest. In front of him, a branch weighted down with thousands of monarch bodies fell off a tree. Butterflies scattered everywhere.

And right there, right in front of him, was a monarch with one of his program's tags.

It seemed too good to be true. Still, some of the most remarkable scientific breakthroughs are coupled with serendipitous events. The discovery of penicillin, for example, was purely accidental.

Chance favors the prepared mind. But this was fantastical, like a Greek tragedy that relied on a deus ex machina to resolve a problem. There may have been as many as half a billion monarchs overwintering in the trees surrounding Urquhart that year. The chances of recovering a tag from his miniscule project were one in twenty mazillion bazillion. Fortunately for Urquhart, a *National Geographic* team accompanied him, so witnesses could attest to the truth of the scientist's fluky good fortune.

Urquhart did *not* discover that monarchs overwinter in a small area of mountains in Mexico; Mexican people already knew that. But he *did* nail down the truth that some of the gathered monarchs had traveled all the way from northern North America.

Urquhart's sighting solved one mystery, but in science, when one question is answered a hundred new questions are raised. The follow-up question was obvious: How did they do it? How did a tiny insect that had never before been to Mexico find its way to these small, special mountain sites where the perfect overwintering microclimate could be found? (Eventually, researchers would learn that the butterflies settled in a number of groves of trees in this high area. A small part of the mountainous region would become a United Nations Biosphere Reserve.)

For decades, no one had any idea. Then scientists began gradually unlocking the molecular secrets of living cells. Today we know enough about how cells operate that we can provide a somewhat complete answer. The story is pretty cool, and it starts, as do most of life's stories on Earth, with the sun.

We are all addicted to the sun. Time ticks in our veins. This is not by choice. It is a biological mandate.

This is the nature of life on our planet. Even eyeless organisms are ruled this way. Our cells pulsate, hour by hour, with the rhythm of dawn and dusk. Even nighttime creatures like moths and bats and hermit crabs and teenage boys are ruled by the twenty-four-hour day.

In attestation of the sun's universal power over earthly life-forms, human cultures have always venerated our golden sphere. The Greeks worshipped Helios. The Aztecs followed Nanahuatzin. The Basques loved Ekhi, the sun goddess who protected human beings. The ancient Australians told a heart-wrenching story of the sun goddess Gnowee, who searched from dawn to dusk across the sky day after day for her lost child, carrying the torch that lit the world. The story was one of pathos: her sadness was the world's blessing. Other cultures told of gods who daily pulled the sun across the sky in horse-drawn chariots, working hard from dawn to dusk to bless humanity with light.

The sun is our eternal "clock"—the conductor who holds the world together, keeping all life in tune in one great symphony. We cannot escape the clock's perpetual timekeeping, even if we hide ourselves away in complete darkness. In scientific experiments, human volunteers who have lived in isolation for weeks still live by a universal internal sun clock.

That we all adhere to a universal clock has, of course, been known since humanity first become capable of thinking. But the biology behind our subjugation has only been scientifically explained within the past several years. The complex revelation was deemed so important that when the phenomenon's molecular details were finally revealed, the researchers earned a Nobel Prize.

Each and every one of our cells pulsates with a perpetual twenty-four-hour feedback loop, a cyclical oscillation of biochemicals that cells manufacture and break down right on schedule each day, coordinating the functions of our entire body. Because of these feedback loops, each cell in our body is coordinated with every other cell in our body—and even with other cells in the exterior world.

We all live by the same rhythm. If for some reason we are not in sync with the life around us—if, for example, we take a jet across many time zones—then we don't feel "right" until our cells tune in with the world around us. If we live where daylight saving time is imposed—Spring forward, Fall back—it takes us several days to get in sync.

This is because within each cell, various genes turn on and off throughout the day, and it takes a while for the genes, courtesy of the sun, to figure out that the time has changed. We need to resynchronize our own internal clocks with the Eternal Clock blazing away in the sky above. This is why travelers who have crossed time zones are told to go outside as quickly as possible after arrival.

This regular flow of activation and deactivation is what researchers mean when they talk about circadian rhythms. I'd always imagined that the use of the word "rhythm" was simply poetic, but now modern microscopes have allowed us to videotape cellular activities. The rhythms—pulsations—are quite real. With the correct technology, we can see the oscillation in real time. The rhythms of our cells look something like a beating heart.

Living by the twenty-four-hour sun cycle is the reason why dogs know to go to the bus stop at three o'clock to pick up their children. Why horses know that 6:00 a.m. is time for the oats to arrive. Why cows come home by themselves for milking at five o'clock. Why babies become restless at around the same time. Why birds head south and return north according to the seasons, and why plants like milkweed bloom and senesce at the same time every year. None of this is accidental. It's all arranged by the Lord of Light, the sun around which we circle.

Even insects are ruled over by the sun. The neuroscientists Russell Foster and Leon Kreitzman, in their book *Circadian Rhythms*, explain that flowers produce nectar at specific times each day, and that insects "know" these times. The floral welcome mat is out at only certain hours. "Bees have a daily appointments book for flower-visiting and they can 'remember' as many as nine appointments a day," they write. "Both bees and plants share a common internal representation of the solar day and they can 'tell' the time and synchronize their internal 'watches.'" So: no living thing on Earth is free.

Including butterflies.

But circadian clocks are not the only synchronizers of life on Planet Earth. Many living things also live by a seasonal clock, called a *circannual clock*. Driven by genetics, this yearly clock ensures that the correct things happen at the right time of year. This synchrony is essential if living things are to survive on a planet that constantly changes. Our world depends upon precision timing.

It doesn't matter whether we wear a watch or not. Bears hibernate in the fall and emerge in the spring. Horses give birth in early spring, right before high-protein young grasses sprout. We respond to decreasing light in the fall by becoming "cozy," by settling into our armchairs, by adopting wintertime behaviors, by going to sleep earlier and waking up later. We respond to increasing daylight with spring behaviors: yearning to be outdoors, waking up earlier, becoming more active.

Monarchs, too, respond to circannual rhythms with specific biological changes. The fall migrating monarch is a different animal from the summertime monarch. They even look different. Monarchs that will migrate emerge from the chrysalis bigger, stronger, and more richly colored than their immediate ancestors. Because they will need to climb high into the atmosphere to soar long distances on powerful winds, the wings of migrating monarchs are shaped to allow for better migratory flight.

Their wings are particularly adapted to ride on rivers of air. Just the way I like to ride rivers in my kayak, monarchs like to ride streams of air. Flight for an insect is one of the most energy-costly things they do. Migrating monarchs offset this by using tailwinds. This ability means that they can fly—"glide" might be a better word—farther since they use less energy per mile.

Their flight patterns differ. Summertime monarchs flit around, skipping from flower to flower for nectar. Males chase females relentlessly, and females try to fuel up with nectar and lay eggs while avoiding masculine aggression. Migrating monarchs are different, directed and focused. They usually do not mate. They have one goal, and one goal only: to get to their destination. On their journey south, they consume as much nectar as possible, putting the sugars and fats away to be used during the cold winter months. They eat so much during the migration south that some arrive at their overwintering destination even heavier than when they began migrating.

Importantly, they become highly social. They are more gregarious, roosting sometimes for only hours but sometimes for days in trees during their flight south, at times so thickly that they roost on top of each other.

"They tolerate each other when they roost," the monarch researcher Patrick Guerra told me. "We don't know whether or not they are actually attracted to each other, or if they are all looking for the same useful site and eventually just end up at the same spot."

I asked: If they are not attracted to each other, why would that be?

"Perhaps at the roost, there is some sort of information sharing," he suggested.

Is this possibly one useful tool that helps them find their way to the correct overwintering site? I wondered.

"Maybe," he answered. "Or maybe they just copy each other." Maybe once they are in Mexico, they follow each other up to the mountains. Or maybe they are all drawn by the same sounds or smells. Guerra and others hope to someday devise a study to answer that question.

One reason why they tolerate each other may be that migrating monarchs do not reproduce. Indeed, usually they *cannot* reproduce, since their reproductive organs are not fully developed. The aggressiveness of the males lessens considerably. Instead of investing energy in the egg-laying process, migrating monarchs divert energy into growing bodies better suited to long flights and to a long overwintering period.

One of the switches that begins this reengineering is—you guessed it—the sun. When the days shorten, the internal clocks of the monarchs note the lack of sunlight and alter development. The insects that emerge from their chrysalises are super-flyers, adept at reading the winds and at soaring great distances.

They are much more driven to fly in a specific direction, Guerra told me: "The fall migrants have this strong behavioral tendency to fly south, while summer monarchs will fly anywhere."

When we earthbound creatures look at the ocean or the sky, we see "water" or "air." But creatures evolved to live in those substrates sense elaborate transportation systems, like our own networks of roadways. Monarchs are somehow aware of these complex systems of air currents and can ride thermals high into the atmosphere to hop aboard them, but the details of exactly how they decipher the winds remain a mystery.

As we talked, the number of questions listed under the category "For Further Research" grew minute by minute. Science is a never-ending story. Once at the end of the nineteenth century, just after the discovery of the electron, a physicist wrote that science's job was complete. Everything there was to discover had been discovered, he claimed. Only a few years later, in 1905, a young Einstein wrote that $E = mc^2$.

So there you are: science will never be "complete" until human beings are no longer curious organisms.

Interested in insects since he was a kid, Guerra started his monarch journey in the research lab of the preeminent neuroscientist Steve

Reppert at the University of Massachusetts Medical School. Reppert, Guerra, and others devoted years to discovering several of the biological mechanisms behind the monarch migration.

It had already been demonstrated that monarchs use the sun as a "compass" to guide them during their flight. The Reppert team repeated that demonstration. Adapting a protocol that had long been in use, the researchers put one of the fall migrants into a large, open-top barrel that had been placed outdoors. The insect was delicately tethered to a post in the center of the barrel.

The only thing the butterfly could see was the sky and the sun above. The design of the tether allowed to butterfly to turn freely and to fly in any direction, although not up or down.

In concert with earlier research, the Reppert team found that these monarchs consistently flew generally southwest..

Non-migrating monarchs did not do this.

"The sun provides this nice directional cue in the sky," Guerra said.

That made sense to me. Humans do the same thing.

Later he added: "What's amazing is that they have a brain the size of a pinhead, and they're doing something that would take us all sorts of complex computations to do." I pictured the vast navigation panel of a jumbo jet, then imagined all of that condensed into the tiny brain of a butterfly.

To test the persistence of the southwesterly flight behavior, researchers reached into the barrel and gently turned each flying fall monarch in another direction. As soon as they released the insects, the butterflies oriented back toward the southwest. Summertime monarchs did not do this.

I watched a video of this phenomenon during one of Reppert's lectures. People gasped. The phenomenon was that obvious. These migrating monarchs are determined little things. Tenacious, unyielding, and dogmatic, they are indeed different from their summer counterparts. They are not to be deterred. They just keep on truckin'.

Behind the remarkable aspect of the butterfly's ability to use the sun as a guide is an obvious obstacle: the sun does not stay still in our sky. From the point of view of us earthly creatures, the sun, as noted by our early sun-worshipping ancestors, seems to travel the firmament.

To account for this apparent movement while keeping a steady course, a butterfly migrating south in early morning will keep the rising sun on its left. A butterfly migrating south in late afternoon will keep the setting sun on its right. At noon, a migrating butterfly will head straight for the sun.

To us humans, this accomplishment seems almost fantastical. How do the butterflies know whether it's morning, noon, or night? How do they know where they should be in relation to the moving sun? How does the butterfly "know" that the sun "moves"? The answer is that the butterfly has an innate timekeeper, the circadian clock, which "keeps time" by producing certain molecules at night which then get broken down over the course of the day.

"During the day, the sun triggers the breakdown of these molecules," Guerra explained. "Light essentially stops the production. This is in sync with our daily light cycle."

"This happens even without the sun. The rhythmicity, the oscillation, will last for about a week. Then it becomes increasingly arrhythmic. Things just become flat and there is no more oscillation," he added. "Additionally, constant light also breaks the clock. You get a flat rhythm. We have been adapted over millions of years to day, night, day, night. . . . If you ever feel out of sync, if you want to get reset, just go camping or go off the grid."

This is true for butterflies, also. And for almost every other living thing on our planet.

After performing a lengthy series of studies, the Reppert team fleshed out more details of how monarch migration happens. In one study, they tricked the insects. They captured migrating butterflies and kept them inside in specialized incubators where researchers could control the "sunlight" by turning an electric light off and on.

Then scientists "shifted" the time by turning the electric light on and off six hours out of sync with the real time. It was as though the butterflies had been flown across six time zones. If this happened to you without you knowing, you would believe it was one particular time, when actually there was a six-hour difference.

You would be confused.

So were the butterflies.

When researchers put these "time-shifted" butterflies outside, in the open-top barrel, where the insects oriented by the real sun, their timing was off. They flew in the wrong direction. For example, their internal clock, set by the electric light, told them that it was 10:00 a.m. and they should keep the sun on their left. In fact, it was later in the day, and they should have been keeping the sun on their right.

"In their little incubator, they think it's one time. Then when you test them outdoors, they still think they're in their little incubators and they behave accordingly. They were using the *wrong* rule," Guerra explained. "When we took them outside, it was morning to them. But it was actually later in the afternoon. They were doing their jobs and following their instructions to the letter—but they were doing this in the wrong context."

But *how* were they doing this? Were they following the sun by using their eyes? That's what we would have done, and would probably have continued doing until the rhythm of the world around us had clued us in on the time change.

Or were they using some other sense—one, perhaps, unavailable to humans? For example, bats navigate by using their own type of sonar called echolocation. Perhaps monarchs also had a special navigation tool kit?

Scientists have long understood that organisms have central governing clocks, along with the circadian clocks in each cell. Guerra suggested: think of a movie where the leader tells everyone on the mission, "Time to synchronize our watches."

"There is One Clock That Rules Them All," he continued, and for us that clock is in a specific region of our brains. Researchers assumed that the One Clock That Rules Them All for butterflies was also in butterflies' brains.

They were wrong.

The Reppert team showed that migrating monarchs fly by using an essential navigation "clock" located not in the brain . . . but in the antennae.

"Monarchs have a central clock in the brain that governs things like the sleep-wake cycle," Guerra elaborated. "But our team found that for orientation, they use clocks that are in the antennae."

Technologically speaking, we do something similar.

"We have multiple clocks all around us that we use for various tasks." Clocks on walls. Laptop clocks. Cell phone clocks.

"But on my workout, I use my wristwatch," he said. "For some reason, we don't know why, the antennal clocks are used for migration. This is why our finding was so quirky and unexpected. You would expect the Clock That Rules Them All would be the clock telling the insect what time of day it was when migrating, but apparently no, they were using another clock, a clock outside the brain."

A butterfly's antenna is where the rubber meets the road, where the living organism encounters the world at large. Their antennae are truly full-service organs. Children sometimes use the word "feelers" to describe antennae, perhaps imagining that butterfly antennae are equivalent to our own outstretched arms feeling our way through a game of Blind Man's Bluff. In an allegorical sense, they are correct.

A butterfly antenna is a natural marvel. Like a Swiss army knife, the antenna is a marvelous multifunctional toolkit with all kinds of tools that perform many essential tasks. It detects odors wafting through the air, sometimes at great distances. It balances the insect in flight. It helps the insect find its way while flying. And it possesses several metaphorical "clocks," including a clock that sends important timing information to the brain.

Guerra and the Reppert team wanted to know more. Butterflies with both antennae navigate very efficiently. When one antenna is removed, the butterfly can still navigate well. But when both antennae are gone, the butterfly can no longer orient and fly south to migrate. Fred Urquhart suggested anecdotally in the 1950s that this might be the case, but the Reppert lab proved it was so, via an ingenious experiment.

Rather than *removing* an antenna, they decided to *paint* both of them. They painted one black, so no light could penetrate. They painted the other with a clear paint, allowing light to pass through.

The insect so painted could no longer orient its flight. Conflicting messages regarding time of day were apparently being sent by each antenna to the brain. They concluded that the butterfly had a timing mechanism that acted as a "clock" in each antenna. When one antenna was absent, the insect could still orient quite well by depending on only the remaining antenna. But when the two antennae were painted differently, the signals sent from each antenna differed. The scientists concluded that the biological mechanisms that track the sun's movement were located in the butterflies' antennae.

When monarchs leave their overwintering sites, beginning in late February, they head north, following the spring's plant rebirth. While a few end up riding winds as far north as the Canadian border and beyond, most stop in Mexico or Texas, where they mate and lay eggs for the next generation. That generation then heads further north. This happens for as many as four or even five generations, until it is time for the late-summer monarchs to head south again.

The Reppert team wondered how the butterflies in Mexico found their way back north at the end of the winter. Some researchers believed that the biological mechanism for this change was triggered by lengthening days, just as the migration south had been partially triggered by shortening days. Others suggested that day length was not the trigger, but that the trigger was instead cold temperatures.

Stakes were high.

A six-pack of Guinness was bet.

To find out, Guerra first captured insects in Texas that were migrating north in the spring. Releasing the butterflies and tracking their flight, he found that the butterflies were also using the sun as a guide, but that their tool kits were now helping them find their way *north*.

The compass had flipped.

How had this happened?

As a next step, the team captured fall migrating monarchs from New

England that were moving in a southerly direction. Keeping them in New England, researchers subjected them to twenty-four days of chilly temperatures, making it seem as though the butterflies had undergone a winter 12,000 feet up in the Mexican mountains.

Placing these cleverly tricked insects in open barrels, researchers found that the monarchs flew north. Even though it was still fall, the butterflies were "re-migrating," as though they had completed the whole migration cycle down to Mexico and back again.

At the same time, they captured and held another group of fall migrating monarchs. They did *not* subject this group to temperature changes. Instead, researchers held them for many months in stable, warm, fall-like conditions. Then they released these insects the following March. The butterflies behaved as though they had been held in suspended animation, which in a sense, they had been.

These butterflies continued flying south, as though it were still fall.

"The overall message is that when we subjected fall migrating butterflies to cold, they flew north instead of south. And when we took other fall migrating butterflies and kept them warm in the lab until spring, they flew south, even while their friends from Mexico were already flying back up north."

"That was the last brick in the house," Guerra said.

They had shown that the crucial cue for migratory orientation was not daylight hours, but cold temperatures.

That's cause for concern, he suggested.

"The whole problem with global warming," Guerra warned, "is that if cold temperatures no longer happen in Mexico, the migrators might never come back north."

The team does not know specific details for these crucial cues. How long do the butterflies need to experience cold in order for their migration tool kit to flip? What is the minimum cold temperature they need? More questions for further study.

To sum up, science has discovered that monarch migration is not a "rote" behavior, but that there are instead a number of environmental cues that butterflies take in and somehow integrate in order to make behavioral decisions. They take into account shortening days, colder and warmer temperatures, growth and senescence of milkweed. Which of these cues takes precedence depends on the environment in which the butterfly's complex behavior plays out. The tendency to migrate—the migration syndrome—doesn't seem to be controlled by any one on-off switch.

Evolution is about variation, and monarch migration is a prime example of that. When it comes to "correct" behavior for migratory monarchs, there is no right or wrong. Instead of these absolutes, there is considerable gray area. There are typical monarch behaviors, but there are also anomalies, butterflies who behave outside the box, who don't conform to the norms. In fact, monarchs seem to be particularly talented at this hedging-their-bets lifestyle.

During the 2017 fall migration, Cathy Fletcher of Santa Barbara, California, was carrying a tray of milkweed plants to her garden when a female monarch arrived. The insect laid five eggs, one on each plant. Then she flew off again.

David James told me this story. I was intrigued. According to what I had read, migrating monarchs supposedly did not reproduce. I called Fletcher for details. We talked in mid-December, when the fires were still raging. Santa Barbara had been the scene of some dreadful events, including the burning of a large racing stable during which many horses had died. Fletcher told me that the fires had not threatened her home, but that she and her husband had been told to stay inside because of the smoke. When they finally were able to go out, they found butterflies lying on the ground, overcome apparently either by the smoke or by ash. Fletcher rehabilitated one—cleaned it off, fed it some nectar, and sent it on its way.

It was much earlier in the migration, in September, that the egg-laying monarch had shown up. "I was just standing there, drinking in the moment," Fletcher told me. She noticed that the egg-laying female had a tag. She took a photo and sent it to David James. The butterfly had been tagged by a volunteer hundreds of miles north in Oregon. It had migrated hundreds of miles, only to wind up laying eggs.

I asked if that meant that the butterfly would therefore not overwinter and learned that this was one of the great unknowns. Would she flip back and forth between the two biological states of winter and summer behavior? Or was she doomed by the egg-laying to thus end her life? It seems unlikely that she could then return to her overwintering biology, but then, it had once also seemed unlikely that northern monarchs would migrate all the way to Mexico.

"It's been shown that the tendency to fly south and the laying of eggs are not as strongly coupled as we used to think," Guerra explained, meaning that under anomalous temperature conditions, a migrating butterfly could have switched into reproductive mode. "The butterfly could have been in the migratory syndrome, but on the way down it may have experienced temperatures that were high enough to encourage reproduction. It seems as if there are two sets of instructions—the migration instructions and the egg-laying instructions." Exactly how these two sets of instructions balance out has yet to be discovered.

So how the butterflies find their way from Canada to Mexico is no longer quite the mystery it once was. "But there are other questions left to answer," Guerra explained. At the beginning of his research career, Guerra has no worries that monarch science is "complete."

He wonders: "How do the monarchs know when to stop? We don't know why they stop in Mexico. There must be something about Mexico potentially that tells the monarchs they are here. Maybe there are cues: We need to stop because it's the right smell. What happens if the forest is gone? Will they still go there? How do they sense the magnetic field? That's all *Star Trek*."

According to accepted wisdom, monarchs behave in very specific ways: A migrating monarch does not reproduce. And yet . . . It was as if, for monarchs, the rules were made to be broken. Perhaps this is the reason why the species, which evolved only a million years ago in a small region in North America, now flies worldwide. Monarchs are specialists not in living in a distinctive setting, like Britain's large blue butterflies, but specialists in adapting to a wide variety of conditions wherever the winds take them, as long as milkweed is there as well.

"Variation is the fodder of evolution," Hugh Dingle, author of *Migration*, explained to me. "Whether or not the monarch goes into reproductive diapause [suspended development of the reproductive organs] depends on the day length. But that can be modified by temperature. The egg-laying is just one more example of variability of monarch behavior. This is especially true in a place like California where the climatic conditions are so variable."

The Golden State may require extremely adaptable life-forms, but Dingle and student Micah Freedman have also studied monarchs in Guam, where the butterflies thrive but don't migrate, and in Australia, where some monarchs migrate but others do not. Monarchs also live on many Pacific Islands, where in general they do not migrate. In other words, within this one species, individual behavior varies considerably.

I took some time to think about this. As a mammal-centric human being, I had never considered that insect behavior would be anything but rote. Simple. Cut-and-dried. But when I thought about it, variability made sense. Butterflies and moths have been around for well over a hundred million years. They wouldn't have lasted so long if they weren't highly adaptable and capable of change. A butterfly makes its living by being out in the world, flying through rivers of air, avoiding predators, seeking out the perfect plant. A monarch's world is even larger if it is part of the long-lived Methuselah generation and must sometimes fly thousands of miles to find suitable winter refuge before returning to lay eggs.

Of course their behavior must be malleable.

The one universal truth Dingle has found is that there is natural variation in every species. When salmon, for example, hatch in the upper reaches of their natal freshwater streams, they swim downstream out into the ocean. Most of them then swim far out to sea. But not all: some stay closer to the shoreline. This is nature hedging her bets. If something happens far out in the ocean to destroy the salmon, there will still be a closer-to-home seed stock to get things up and running again.

Monarchs are similar. The *tendency* to migrate has been found in monarch lineages, implying that there is a genetic component, but that tendency does not predict with any certainty whether or not any *individual* monarch will migrate. In Florida, for example, where monarchs can survive the winter, some monarchs migrate while others do not. The difference is most likely the environmental conditions in which the insect lives.

Guerra thinks of insects as "walking sensors" that must respond to a variety of the planet's dynamic conditions. "This makes sense because you cannot predict what's going to happen. Monarchs are always trying to be in the 'Goldilocks' zone, where everything is just right."

That's why they have evolved the ability to migrate—but don't *have to* migrate.

Hugh Dingle and colleague Micah Freedman wanted to know more about the difference between migrating and non-migrating butterflies. Was non-migration due to a loss of the biological *ability* to migrate, or was it due to the loss of the proper environmental cues? They captured Australian monarchs that did not migrate and bred them with each other under conditions of decreasing daylight.

But then they found something remarkable.

"The transition to residency may not be irreversible," Freedman told me.

I found that interesting, but what I found even more interesting is that they suggest that this lifestyle decision occurs not during the butterfly stage, but during the *caterpillar* stage. When Freedman and Dingle subjected the caterpillar offspring to fall-like conditions, they found that the caterpillars spent several extra days in the caterpillar stage, eating like crazy and packing on the protein and fat, as though they were preparing for a long flight.

So it turns out that how caterpillars experience their world may determine what the butterfly will do. In a fascinating and ingenious experiment, the evolutionary ecologist Martha Weiss has suggested that this might be true. She subjected caterpillars that would become moths to a certain aroma that was coupled with an electric shock. Most of the caterpillars learned to avoid that aroma. When the moths emerged, she found that most of the moths also avoided that aroma. In other words, information gathered by the caterpillar carried over through the transformation stage and integrated into the behavior of the adult.

This has been described as "memory," but biologists do not use this term in the human sense of remembering what we ate for breakfast. It's more like that the "memory" is a biologically encoded tendency to avoid that carries over into the butterfly stage. This entire line of research—studying how information gleaned during the caterpillar stage benefits a flying insect—is only just beginning to be explored. It's likely to yield intriguing results that will help us better understand how experience and biology interact during our own formative years to create the adult personality.

From our point of view, we see the butterfly in flight as having "broken free" of its earthly chains. But it turns out that nothing could be further from the truth. The "new" organism, the butterfly, is a compendium of worldly events experienced by the caterpillar.

We also tend to think of monarch migration as something highly unusual in the butterfly world. That, too, turns out to be wrong. Lots of butterflies migrate.

On the border between China and Pakistan, the Himalayan mountain known in the West as K2 is only the second-highest point on the planet (Everest is first), but it is perhaps the most inaccessible. Few climbers attempt its ascent, but of those who do, one out of four dies. Hence the peak's other common name: Savage Mountain.

At 28,251 feet, the summit stands out above the surrounding range, looking like a pyramid, but with crazy, vicious angles. Technically formidable, it is often shrouded in clouds. Violent storms last for days. Winter comes early. Summer is little more than the blink of an eye. You have to start out at just the right time of year, and even then you cannot dawdle.

On July 30, 1978, Rick Ridgeway, one of the few climbers to successfully reach the peak and survive the descent, was just beginning his climb in the company of a large team. They'd had to start a bit later in the season than they'd wanted, and then encountered storms that impeded their progress. Hovering like a predator waiting to strike, discouragement stalked them.

They'd been climbing for days, but were still at "only" 22,000 feet. It was just about noon. At least the skies had cleared and the sun was out.

"My mind drifted, hypnotized by sharp colors and thin air and warm sun," Ridgeway wrote in *The Last Step*.

Then, in that world of white and rock, his eyes took note of inexplicable bits of color fluttering overhead like shards of stained glass.

"A butterfly landed near the rope. It was a beautiful butterfly, about three inches across, piebald orange and black like the painted lady butterfly back home."

The sight shook him.

"A butterfly? At twenty-two thousand feet?"

He began counting and quit when he got to thirty.

". . . a cloud of them flying up from some unknown place in China, rising on the air currents up the mountain ridge."

Was that even possible? he wondered. Was it just a figment of an oxygen-deprived brain? The team took photos, in case no one else believed them.

The experience was life-changing for Ridgeway. Whenever someone asked him why he went through the pain and discouragement of climbing, he told them about the painted ladies.

But he wondered, "What could have been the reason for this lemming migration?"

Forty years later, the Spanish evolutionary biologist Gerard Talavera has something of an answer for him. Talavera, a lepidopterist based in Barcelona, is also a mountain climber. When he heard about Ridgeway's experience, he was impressed. But he was also a bit skeptical. He wanted proof. The climbing team provided him with the photos.

"It's the highest record of any insect flying free that's ever been reported," he told me. "Obviously, it cannot fly much higher than that."

By this he meant that if the butterflies flew any higher, there would be no real atmosphere to speak of. These butterflies ride the updrafts of air currents that normally occur because the air at the base of mountain chains is usually warmer than higher up. The warm air rises. Painted ladies (and plenty of other living things) take advantage of the lift, in a sense hitchhiking over the mountains by getting this free ride.

The painted lady is a phenomenal butterfly. Painted ladies live in almost all parts of the world, but differ slightly on each continent. Her wingspan is only about half that of the monarch, but she migrates equally long distances. In fact, sometimes her migration distances are longer.

She has a migration pattern that's somewhat similar to that of the monarch. When the weather begins to turn, if she is north of the Arctic Circle, she begins a migration that can be as long as 2,500 miles, and she may accomplish that task in only a week. If she is flying from Europe she ends up in sub-Saharan Africa just at the end of the rainy season, when there's plenty of cover for laying eggs and plenty of plants to feed on. She is more flexible than the monarch, in that she can take advantage of a number of different plant species.

When the Sahel dries up in the spring, the offspring of the painted ladies that flew south in the fall will ride the northerly breezes to return to Europe just in time to enjoy the newly emerged greenery. Unlike the monarch butterflies, the individual painted ladies that made the trip south do not survive long enough to fly back north.

In Europe during the spring migration, clouds of the butterflies are so commonly seen that, says Talavera, "it's a phenomenon that even people not familiar with butterflies notice." The migration south in the fall is less visible and for quite a while it was assumed that the fall migration either did not occur at all, or only occurred sporadically. But Talavera and others have found that the fall migration is a dependable pattern. The cause of the misunderstanding is that the insects fly so high in the fall that people normally do not see them. Riding on high thermals, they soar south over the whole of Europe, over the Alps, over the Mediterranean, across the Sahara, and then land in the sub-Sahara, where grasses and bush provide them with food and cover for breeding and egg-laying.

"The same butterfly makes the whole trip to Africa," Talavera explained, "but it is not the same butterfly that makes the trip back in the spring."

There is an important difference between painted ladies and monarchs. Monarchs overwinter on high mountains and do not breed during that time.

Butterfly migration is not at all uncommon. In *Migration*, Hugh

Dingle tells of seeing a mass migration of an Australian butterfly called the Caper white. Looking out the window of a fifth-floor apartment in Brisbane, he counted 48,000 to 52,000 passing insects per hour, for two and a half hours. They seemed to be riding on the wind and were so intent on reaching their goal that they bypassed many flower-rich gardens. "No butterfly even hesitated at profusely flowering bushes," he wrote, "although there were butterflies of other species nectaring at them."

How do you "count" a river of butterflies?

"I made ten one-minute counts in 30 minutes of butterflies going through the garden of our building, which was about thirty meters wide between the building and the river," he told me. "These averaged 82.2 butterflies per minute, so that was about 822 × 30 = about 24,660 butterflies in 30 minutes. I could detect no obvious change in the density of butterflies over the next two hours—so 24,660 × 2 = about 48,000 to 52,000 butterflies per hour for 2.5 hours."

This was useful information, because I was about to encounter other astonishing estimates of insect populations. Aided by recent technological improvements, Jason Chapman, who calls himself a movement ecologist, estimates that annually about 3.5 trillion insects cross the skies of Britain annually. Others estimate that as many as 4 to 6 billion dragonflies migrate from Europe to Africa at one time.

There are a whole lot more of them than there are of us. People don't notice them because they are usually too high in the sky to be visible to the human eye. That may be one reason why their populations continue to be high during certain years of heavy rainfall. Their nondescript appearance may well be their saving grace. While rare butterflies continue to be collected and traded, sometimes on the black market, almost no one covets the ubiquitous painted lady.

Thirteen

PAROXYSMS OF ECSTASY

The eyes of butterflies are remarkable, because they are nearly as
diverse as the colors of wings.

Adriana Briscoe

Butterfly thievery never truly disappeared. Even now, in the twenty-
first century, it's alive and well and in the headlines worldwide.
National Geographic's August 2018 issue ran an article about butterfly
smuggling: "The trade in rare butterflies—both legal and illegal—spans
the planet."

The article featured a butterfly catcher who, like Victorians of old,
risked his life to climb treacherous high cliffs to net Lepidoptera that in
today's world would ultimately fetch thousands of dollars. The human
passion for butterflies is so strong and the business still so lucrative that
some people continue to collect, despite international bans on trade in
some species. A scientist who lived in Las Vegas once beckoned me
surreptitiously into a back room in his home. He unlocked a padlocked
door. He showed me drawer after drawer after drawer of pinned butterflies, each more beautiful than the last. Many were illegal to possess.

In 2007 Hisayoshi Kojima, who claimed to be "the world's most
wanted butterfly smuggler," was convicted of playing the global black
market when he tried to sell a U.S. federal agent a collection worth a
quarter million dollars. He went to jail. But I wouldn't be at all surprised

if he continued to trade and collect. Shades of sad old Herman Strecker. Of the Rothschilds, Lord Walter and Dame Miriam. Of Bates and Wallace and of all the lepidopterists who for hundreds of years have been beguiled by the butterflies' seductive wing colors. Like Maria Sibylla Merian, some people find butterflies impossible to resist.

The penchant is innate, wired into the complex information pathways that course through the human brain. Think of Konstantin Kornev's young daughters out in their South Carolina field on a summer day. Or think of non–*Homo sapiens*: the male Vogelkop bowerbird—that fabulous avian architect who builds stupendously complex nuptial palaces to lure females impressed by artistry—chooses bits of butterfly wings to adorn the walkway to his structure.

Response to color has been embedded in neuronal pathways ever since complex animals first began to evolve in the planet's oceans, all the way back to the 540-million-year-old beginning of the Cambrian Period, if not earlier. Hence, butterfly wing color is bewitching, enticing, intoxicating, demanding, driving, forceful, compelling, and downright sexy.

At its most basic, visual beauty—or any other kind of beauty, for that matter—is neural excitation. It's also a lot of other things, of course, like a lifetime of learning and experience and ideals and cultural influence. But at its heart it's an intense engagement of the receiving organism's brain with something important, even essential, in the world outside that brain.

Here's how all that jargon works in real life:

Imagine that you drive by the same apple orchard once a month throughout the year. It's a fairly humdrum experience until, one day in September, you drive by and see the orchard's hundred trees all decked out in red: the apples are ripe. You do a double-take. You are immensely impressed by the orchard's sudden beauty. This is a universal human response. How many billions of pictures have kids around the world drawn with red apples standing out against the green leaves of a tree? Wherever there are trees with ripe apples, there are likely kids drawing them.

Next you put your foot on the brake and assess the possibility of getting some of those fresh-off-the-tree apples to eat. The vision of red has reminded you of earlier apple experiences, and you start salivating at the memory of the fresh, cold juice. It's visceral. Inextricably linked to survival.

Here's what happened: the color red hijacked your psyche.

We are fortunate to be able to experience this. Our distant primate ancestors could not. They possessed only two kinds of color photoreceptor cells, or *cones*: for blue and for green.

But around 30 million years ago, our line of primates evolved a third color cone. This might seem nearly miraculous, but it was brought about by a fairly simple event known as gene duplication. When we evolved a third color cone, coupled with the blue and the green color cones, a whole new glamorous, glorious, glowing world of bright reds and radiant oranges and cheerful yellows opened up to us.

Along with those joyous colors came a propensity to detect and grab ripe fruit more easily. Research suggests that this need to reach out for the desirable—the *beautiful*—is wired into our psyches. Studies show that the reward/pleasure center in the brain activates when we see something we think of as beautiful. The same happens when we enjoy delicious fruit. We want to grab that beauty, sometimes quite literally: eat the apple.

An entirely different center in the brain activates when we see something ugly: our muscles prepare to escape. We want to run away, at least on some level.

A just-emerging science called *neuroesthetics* is discovering how our brains respond neurologically to beauty. Asserting that the foundation of beauty is about survival, neuroesthetics is deeply rooted in evolution: We are attracted to things that aid us to survive. In this view, beauty is sensory exploitation. It takes advantage, suggests Michael Ryan in *Taste for the Beautiful*, of our already present "hidden preferences."

We are usually unaware of these hidden preferences. One study I love addresses the universality of a particular natural scene: a grass-covered plain with maybe a tree or two, flowing water, and a hillside or cliff or even a mountain. Turns out that the observer usually imagines himself not on the plain, and not in the water, but up on the cliffside, looking out over the scene. According to the research, viewers of many nationalities and cultures worldwide rated this kind of landscape as their favorite.

They often connected the scene with words like "peaceful" and "serene." In other words, "safe." This view of beauty has evolutionary roots. Floating down the Save River in Zimbabwe with companions one long-ago evening, I heard a raucous, cacophonous, nearly deafening screeching. Truly terrifying. Hundreds of baboons were clambering over an escarpment high above us, then settling on various perches where they would spend the night. There it was, right in front of my eyes: the universal landscape scene. The hidden preference was chosen this time not by *Homo sapiens*, but by baboons, who could thus sleep peacefully without fear of being attacked by lions or hyenas or wild dogs.

So beauty is *not* in the *eye* of the beholder.

It is in the brain's reward system.

For many living things, it exists within the context of the brain's visual processing system. Here's how it works in a nutshell: When our eyes open in the morning, we take in light and we "see" the world around us. Photons of light enter the eye and activate three different kinds of color-detecting units—"cones"—located primarily in the center of the retina. We note that the sky is blue, that the spring grass is turning a delicate green, that last night's red shirt lies in a heap on the floor.

That information is transmitted through the optic nerve, along a defined pathway in the brain. Via several centers, the visual message goes from the front of the brain (the eye) all the way to the back of the brain, to the primary visual cortex. Here the information is sorted and switched to different tracks. Information about color travels along a pathway that runs along the base of the brain. Information about movement runs along a pathway that leads to the top of the brain.

Weird, isn't it?

The first pathway is called the *inferior pathway*. The second pathway is called the *superior pathway*. So when you see an apple hanging on a branch, swaying in the wind, your brain is thinking about this in at least two entirely separate ways. No one yet understands how these two pathways become integrated again. By the time the information transforms into conscious thought, you say to yourself: There's an apple swaying in the wind.

Your brain processes color information much, much more quickly than the information about movement. The difference in processing time is astronomical—to the gazillionth power. What that means is that the *color* of an apple—or, in a spillover effect, the color of a butterfly—hits us fast and hard, in the gut.

The language of butterflies is the language of color. In an evolutionary sense, butterflies really do *intend* to be jaw-droppingly beautiful (although not consciously). They don't intend to impress us humans, of course, but since the language of color is both primal and universal, we are impressed anyway.

A quick primer: There are many different kinds of eyes in the animal world. Not all eyes are like our human eyes, which we call camera eyes. But all eyes do have one thing in common: an eye is a survival tool, and it will have evolved specifically to help the organism not to see the world as it "really" is, but to survive in a universe filled with danger. Eyes are there to help us eat and not be eaten, and to help us find mates.

The earliest "eyes" were simple sets of cells on the surface of a living organism that responded to light. If you were living in the ocean (which all life-forms then were), this would help you determine up from down. "Up" would be toward the light. "Down" would be away from the light.

Eyes eventually became more sophisticated. Their evolution depended entirely on the lifestyle of the organism. Where did the organism live? What did it need to survive? Who were its predators? How did it eat? Eyes became so important that a major event—the Cambrian Explosion of 540 million years ago, in which countless new organisms evolved in the world's oceans—is attributed to new developments in eye evolution: the more you saw, the safer you were. If you were a predator, you needed one kind of eye. If you were prey, you needed another.

Fast-forward hundreds of millions of years to the fabulous eyes of butterflies. Unsurprisingly, our day-flying insect friends have staggeringly sophisticated eyes—eyes that are particularly expert at perceiving and responding to a myriad of colors created so gloriously via light from the sun.

With only three color cones or "channels," our eyes suffer from an information bottleneck. We have sacrificed the ability to see multitudinous colors so that we can see with considerable 20/20 acuity. Butterflies chose a different path. We would think of their vision as blurry. However, some butterflies have six, seven, eight, or even more color channels, and their world is filled with a riot of color.

Butterflies have compound eyes rather than camera eyes. Compound eyes have lots of "little eyes" within the complete eye. These "little eyes," called *ommatidia*, are ordered within the complete eye structure in highly organized rows. A rough analogy is to think of ommatidia as similar to the pixels that make up a picture in a newspaper. Because of this, researchers suggest that butterflies may not assemble a picture of the world around them as our brains do, but instead may see the world as a kind of rough "mosaic" of color. For us, detecting the lines and edges of an object is essential. We have cells in our brains that respond specifically to vertical lines in the world, and cells that respond specifically to horizontal lines.

Now here's where things get really weird and extremely interesting: each ommatidium in each butterfly eye has its own set of tools for perceiving colors and other important information. So one row of ommatidia in a compound eye may respond to the presence of one particular color, while another section of the same compound eye—another group of ommatidia—responds to other colors.

In some species, color vision is even more marvelous. The common cabbage white, barely noticed by even avid butterfly collectors, has eight different types of photoreceptors. They are not all used to detect color in the conventional sense that we usually mean. Specific wavelengths of blue trigger feeding reactions in this insect, and when the cabbage white female detects a specific wavelength of green, she responds with egg-laying behavior.

No one yet knows how these various sensitivities are integrated in the brain of the cabbage white butterfly or even if they are integrated. The butterflies' responses to the color around them seem to be highly stereotyped, and they may simply have no choice about how to respond.

Other butterflies, however, have been shown to be quite capable of learning and of modifying their responses to colors. Not surprisingly, monarchs reside in this group. Given that their tasks in life require considerable decision making rather than simple stereotyped or rote behavior, this makes sense. Any living thing that's capable of traveling hundreds of miles over numerous ecosystems in only a few days must be able to learn and change behavior.

The biologist Douglas Blackiston, the entomologist Adriana Briscoe, and several other colleagues tested monarchs, first to probe the details of their ability to see color and then to discover if the monarchs had innate color preferences that could be modified through learning. They found that monarchs really, really, really like orange. No surprise there. They like yellow, but only half as much as orange. Blue, not so much. And—surprising to me, at least—red, even less.

Next, the scientists trained their monarchs to go to different colors in order to find a sugar reward. The sugar rewards were associated with yellow and blue and red colors, the colors that they hadn't been interested in. Most of the butterflies caught on immediately. They even got the butterflies to associate the color green with sugar. This was rather unexpected because in real life, green indicates leaves rather than nectar.

When I first heard the story of Amelia's persistent butterfly, it struck me that monarchs had to be particularly intelligent.

I asked Blackiston what he thought.

"Everyone thinks that bees are the geniuses of the insect world, but for me the female monarch is the epitome. The female is the essence of a single working mother. A monarch born in Boston will migrate all by herself to Mexico. I don't know even with my GPS how easy it would be to find Mexico from here."

Intelligence, he said, is a prime characteristic of monarch butterflies.

"If you are in Boston, your food plants are what's around Boston. But when you get to North Carolina and then to Mexico, your food is going to be very different. How do you know how to do that?"

The he answered his own question: "You build a brain that allows you to learn."

Blackiston and his colleagues wanted to know how quickly the butterflies could learn. What kinds of cues did they pay attention to?

They created artificial flowers with food rewards. Each flower was painted a different color. They released the butterflies and found that the butterflies learned quickly to go to many different colors, depending on where they had learned that they would find food.

"Monarchs have very robust learning capabilities for a simple, small insect. They're actually incredibly interesting and smart creatures. Training frogs is way harder."

"They are exceptionally capable of learning new things," Blackiston concluded. "The big thing we were thinking about is that there has been a lot of disruption to the monarchs' migration routes."

I thought of Amelia's butterfly, navigating the Willamette Valley that has changed so much in only the last century.

"It's been critical to understand how robust they are, how much learning these animals have. If they're hard-wired, they could easily die. It turns out that they are quite clever. In fact, there are now so many ships out in the Gulf of Mexico that they are using a whole new route—hopping from ship to ship out to sea."

I checked. There are indeed scores of photos available of monarchs using ships and oil derricks as resting places as they head to the Mexican mountains. The jury is still out, however, on whether using oil derricks as rest stops is a good thing or not.

I decided to learn more about how North American monarchs behave during their fall flight south by literally following their journey from near the Canadian border all the way to their favored mountains in Mexico. As noted earlier, millions upon millions of monarchs head south each fall, beginning their migration in late August first one by one, then in small groups, then joining up into clouds of migrating insects that seem, by the time they fly across the U.S.-Mexican border, to be a veritable river of flying colors, glittering in the sunshine.

Or, at least, that's the way it used to be.

Fourteen

THE BUTTERFLY HIGHWAY

To plant a garden is to believe in tomorrow.

Audrey Hepburn

On a late-August day in 2018 I was sitting on a bench in the University of Wisconsin's butterfly-friendly arboretum in Madison. The weather was picture-book perfect. Temperature a fine, dry, obliging 73 degrees. Sky crystal clear. Visibility seemingly infinite. Primates were born for such days.

The slightest of breezes wafted through the waxy leaves of the bur oaks scattered around me. The birds were out for a little late lunch. The bees were busy storing honey and the sounds of crickets noted the shortening days. Contentment. As I jotted down thoughts, Lepidoptera drifted by in the afternoon sun. Swallowtail butterflies with just a hint of luminous structural blue on their wings enjoyed the purple blooms of a tall thistle. Monarchs fluttered everywhere, absorbing nutrition through their proboscises to store in their abdomens, preparing for their long journey south. They were already joining up with each other, socializing and roosting, waiting for just the right winds to send them south to Mexico.

I felt like I was in the middle of a 1930s Disney animation, complete with whistling birds and silly music: Walt Whitman's "butterfly goodtime." Listening to the rustling sounds of tall grasses, enjoying a sun that was hot but not too hot, I couldn't think of a thing to complain about.

Very odd for me. I thought about worrying about what was wrong with me for not feeling worried, but then decided not to bother even with that. "Let's go get drunk on light," post-Impressionist Georges Seurat once wrote. I knew exactly what he'd meant. I'd pigged out on so much sunshine that I could barely move. I'd rolled a Lucky Seven.

Unfortunately for the fair city of Madison, just one day earlier Kane County had been drowned in a deluge that would have impressed even Noah. The storms pounded the land with as much as eighteen inches of rain in one twenty-four-hour period. Sadly, one poor fellow had been washed away by an unexpectedly strong current.

The city's infrastructure couldn't keep up. A woman in an airport line told me her family had had to evacuate their home, not because of the water itself—but because Madison's sewers were flowing *backward*. Sewage was bubbling up into her basement.

Courtesy of global weather changes, rising lake levels had inundated the isthmus on which Madison is built. Yet another storm would arrive tomorrow. Fortunately, I would be on a flight out very early the next morning. I would be the rat abandoning the ship.

I had come to meet the arboretum's new director, Karen Oberhauser, the U.S. grand dame of monarch research and creator of influential classroom education projects on monarchs. Oberhauser had just left her longtime position heading up the Monarch Lab at the University of Minnesota. A protégée of Lincoln Brower, she has been involved with monarch research for most of her career and sits on the board of Monarch Joint Venture, a collaborative group of scientists working to improve monarch numbers. Because of her dedication, the Obama White House named Oberhauser a "Champion of Change."

It's not surprising, then, that she had arrived here, in this nearly century-old institution, as a change agent. The Madison arboretum, founded to be a showcase for Wisconsin's various ecosystems, had not previously emphasized monarch conservation. But after only a few months under Oberhauser's directorship, it was already obvious that this would change. The arboretum had just become the nation's first such facility to join the Joint Venture program. The visitor center had plenty of information on butterfly conservation, and all you had to do was step out the door to see plenty of monarchs in action, fueling their abdomens with nectar for the journey south. Soon other monarch-oriented professionals would join Oberhauser.

During our visit, we walked over several parts of the 1,200-acre conservation and research property, observing the plants and noting the sometimes severe damage resulting from the unprecedented torrents. We looked at a section of roadway that bordered one of the arboretum's small ponds. "Used to border" would be a more accurate statement. A large chunk of road was now gone. As the rains continued through that season, even more chunks would disappear.

Oberhauser was worried about the effect of all this water on the arboretum. And yet, not despite but *because of* these deluges that had pummeled large parts of the Midwest and East Coast, it was already obvious that the 2018 monarch migration was going to be the best it had been in many years. At least in the Midwest, the climate anomalies had been good for plants. Growth was spectacular. This meant more flowers for nectaring, which meant better-fed insects, which meant more productive butterfly matings, which meant more caterpillars. . . .

It also meant that there would be better stopover habitat for monarchs heading south. Of North America's three major flyways—western, eastern, and central—the most significant is the central flyway. From north of the Canadian border, from as far west as the eastern front of the Rockies several thousand miles east to the Appalachians, the central migration route resembles a colossal oil funnel that encompasses roughly two-thirds of the continent.

When it's time for the migration to begin, first in twos and threes and then in tens and twenties, and ultimately in the thousands, monarchs become social, as I've mentioned. In the center of the continent, on the northern shores of the Great Lakes, even as Oberhauser and I walked and talked, these roosts were popping up like flash mobs just before dark on August evenings—only to vanish the next day by 10:00 a.m. or so.

Some of these 2018 roosting events in Canada turned into quite a human hoopla. When word got around that some of the roosts had hundreds of butterflies, hundreds of people turned up to see the spectacle. Quite a party. And then the insects vanished, soaring away when the winds and temperatures improved.

By September 5, only a few days after my visit to Oberhauser, at least some of them had successfully crossed Lake Erie. This was common knowledge because sightings of numerous monarchs were noted and photographed by observers in Erie, Pennsylvania. Their citizen scientist findings were posted on Journey North, a website that began in 1994 by tracking migrations of spring monarchs from Mexico into the U.S. and that has since expanded to track the journey south as well, funded by the Annenberg Foundation.

It's run by Elizabeth Howard, who wanted to explore how the internet could be used to encourage conservation and citizen involvement. Since then, it has grown exponentially and now includes thousands of participants, who use their cell phones to photograph individual monarchs as well as roosts. They then post that data on a Journey North map along with relevant comments. Thus anyone who checks out the website can follow the migration of the monarchs north in the spring and south in the fall.

When I spoke with Howard on the phone during the early weeks of the 2018 migration, she gushed with enthusiasm.

"This has been the most exciting year, in years and years and years," she told me. "Everybody across the breeding range has been talking about how productive the numbers are in their local areas. Everything points to a really positive season. The numbers this year are at least fourfold above last year."

I asked: Why has this been such a great year?

"We saw the earliest-ever beginning of the breeding season. The monarchs returned early in the spring. And we started seeing them in June in numbers we wouldn't normally see until July. Since then, the numbers just kept building. Now we are a whole generation beyond what we normally get."

Their progress south is never a steady trek. Whenever possible, they must stop for food. One Journey North observer reported a previously tagged male who had hung around in Canada, on a lake's northern shore, for a few days. He had been fueling up on nectar from blooming fall flowers. Over the course of one week, from the first tagging to a later recapture, monitors found that he had increased his weight by more than 50 percent. That's how important nectar plants are on the butterfly highway leading south.

A second male, tagged at 10:00 a.m. and then caught again four hours later at 2:00 p.m., had increased his weight by 34 percent. What *had* he been eating? Apparently he had found the butterfly equivalent of extreme chocolate cake—double chocolate with buttercream frosting. Maybe with some ice cream on top.

Migrating monarchs pig out like this for one reason: they need fuel. Surfing the winds, lovely as it sounds, requires a lot of energy. Best to eat as much as possible when food is available. There's another reason. If they arrive at their overwintering site in the Mexican mountains depleted of energy reserves, they may well not survive their winter ordeal, huddling against the cold while undergoing unavoidable fasting. In the Michoacán mountains, inside the Monarch Butterfly Biosphere Reserve at 12,000 feet, the insects will find little, if anything, to eat. They have to survive until at least February before starting their return journey north.

And so monarch advocates began their "butterfly highway" project. All along the central flyway, state and local governments, gardeners, farmers, property owners—whoever they could entice—would be encouraged to plant a plethora of nectar sources. Some of these plantings, they hoped, would be various milkweed species. This was to provide egg-laying opportunities for female monarchs heading *north* in the spring. But for the journey south, a variety of native plants would do: joe-pye weed, various species of goldenrod, butterfly bush and butterfly weed, verbena, asters . . . The list is quite long. Whereas milkweed and *only* milkweed will do for egg-laying, nectaring monarchs use many different plants.

• • •

In her office, Oberhauser and I talked about the numbers of migrating monarchs this year. Her excitement was infectious.

"We're seeing a lot this year," she said. "If they have a good fall migration, they'll do well."

But she was still cautious. Even if this spectacular summer brought about a fabulous southern migration resulting in huge numbers of monarchs in Mexico, to her mind that didn't guarantee the future of the iconic butterfly.

She said: "Populations bounce around. A lot."

Although early-migration anecdotal evidence was encouraging, she warned, until the monarchs arrive at their winter Mexican destination, there is no definitive way to estimate their population numbers. Unlike the small blue butterflies, monarchs don't really have a home base. So the best way to estimate yearly populations is to estimate the size of the overwintering roosts in the mountains of Mexico.

Figures are presented in terms of hectares of trees occupied in the region, but even that is only an estimate, since we now know that roosting monarchs don't stay put for the whole of the wintering season. Nevertheless, this per-hectare data is the most definitive scientists have.

Records have been kept this way since the winter of 1994–1995. In 1996–1997, almost 21 hectares were occupied. But the next year only 5.77 hectares were occupied. A drop of 75 percent.

That in itself wasn't necessarily cause for alarm, since the species' numbers are highly volatile, as bouncy as a super-rubber ball. For insects, extreme differences from year to year are the norm, rather than the exception. But over the twenty-five or so years that per-hectare records have been kept at the Mexican overwintering sites, despite the bouncing-ball numbers, there has been a clear downward slope. A crisis point was reached by the winter of 2013–2014. Numbers had dropped to a near-disastrous 0.67 occupied hectares.

When numbers are that small, one severe climate event could wipe out nearly the whole of the central flyway population. Something like that has already happened. During the 2015 fall migration, Hurricane Patricia barreled toward Mexico at precisely the time when the butterflies were moving toward the mountains. The two trajectories seemed likely to intersect.

When Patricia's winds reached 215 miles per hour, residents and tourists both began to flee. Monarch advocates were on tenterhooks. "How does an insect the size of a paper clip fare in a hurricane?" asked one Mexican newspaper, echoing the fears of many monarch fans. But when Patricia hit land on the nation's west coast, the storm dissipated. Simultaneously, the butterflies themselves shifted their migratory path. They seemed to sense the coming weather. They may have taken refuge in ravines and other sheltering areas of the Sierra Madre Oriental.

Another weather disaster in January of 2002 was not so easily escaped. Normally the overwintering region is dry at this time of year, but on that occasion rain began falling. At higher elevations—where the butterflies are—the rain turned to snow. For three consecutive nights, temperatures fell below 30 degrees Fahrenheit. Clustering together in the Mexican mountain forests helps the butterflies keep warm, but these temperatures were too much for the ectothermic insects. Observers began seeing them fall from their branches onto the ground, where they lay either cold-shocked with wings in tatters, or dead. Scientists believe that the insects might have survived the cold temperatures had they not already been wet from the rain and snow. Moisture and freezing temperatures had delivered a one-two punch.

In early October 2018, I walked with volunteers looking for monarchs and other butterflies at one of the most unusual stops along the butterfly highway that I expect ever to see. I had come to The Wilds, a nonprofit site in southeastern Ohio that bills itself as a safari park, conservation center, and living laboratory.

At The Wilds you can take a "safari ride," a narrated bus tour that takes visitors past the variety of exotic animals who roam this park of nearly 10,000 acres. The site harbors endangered animals such as Grevy's zebras, white rhinos, onagers, scimitar-horned oryx, and even Przewalski's horses. For an extra fee, you can get a behind-the-scenes tour and meet the animals' keepers. You can also go on a horseback trail ride, take a zipline tour, go fishing, stay overnight in a yurt, go hiking and cycling.

The Wilds hosts a remarkable restored butterfly habitat. Since 2004, volunteers and staff at The Wilds have regularly walked the same transect line over and over again. As I walked this line with them, we looked for butterfly species along the line itself and up to fifteen feet on either side of the line. When someone saw a butterfly, they called out and the record keeper wrote the sighting down.

"Monarch," someone called out.

"Oh, that's a beauty," another said.

And it was. Its wings were a rich orange, so much so that the insect was almost red. It looked new and fresh, as though it had emerged only a day or so before. This was quite possible, as the area had been planted with swamp milkweed many years ago, and the species continues to thrive here.

The monarch flew from flower to flower, fueling up for the trip south. There was plenty to imbibe. Several species of goldenrod were ubiquitous. There were purple asters and small white asters nestled comfortably among the tall grasses. A few end-of-season coneflowers remained. A limited number of milkweeds continued to bloom, and there were plenty of open pods, which promised a good crop the following year.

We were walking in a butterfly Nirvana. We saw plenty, including cabbage whites, of course, and viceroys and skippers and yellow sulfurs and eastern tailed blues, cousins of the Karner blues.

There was a lot going on in this species-rich prairie. I had been told to wear sturdy shoes and arrived in a pair of sturdy but low ankle boots. My host Rebecca Swab, director of restoration ecology, took one look at them and shook her head. Fortunately, I keep all kinds of footwear in my car, since you never know what kind of adventure you might find in any one day. There are usually kayaking shoes, flip-flops, riding boots, athletic shoes. . . .

On that day, my preparedness paid off. I pulled out a pair of heavy leather lace-up boots, the kind for heading into the back of beyond. I felt sure she would think this was overkill. Instead, she nodded approvingly.

It didn't take long to find out why I needed them. This was a short walk of less than a mile, but much of it involved mud and out-of-control streams. Recently beavers had found out about the place. They had been busy. We left the "civilized" trail, nice for casual tourists, and walked up into a small bit of forest, then down into a wetland area. The ground was littered with beaver leftovers: wood chips, tree stumps, and half-gnawed branches. With the help of nature's super-talented engineers—and with the help of the endless rains that had fallen on the region throughout the summer—the wetlands had far exceeded their known boundaries.

We crossed a stream on what had once been a minimal footbridge but was now mostly submerged logs. We trudged through mud. Beavers had taken over large parts of this particular butterfly habitat, but that wasn't a bad thing. Frogs were croaking. There was horse balm everywhere. Queen Anne's lace flourished. In the end, this evolving system would provide even more nurture for Lepidoptera.

This nature "park" was thriving. Life here had decided to do its own thing and did not want to be contained by any false, human-engendered limits. So what if human beings wanted the stream and pond in one particular place? The beavers had other ideas. The Wilds staff let the beaver-driven natural succession choose its own course. All this was terrific for the butterflies. Flowering plants grew everywhere along the edges of the mess made by the beavers.

But frankly, none of this should have been there. This whole park of nearly 10,000 acres is a former strip mine. I am all too familiar with strip mining. I grew up in southwestern Pennsylvania during strip mining's heyday, when the mine owners could—and did—do as they pleased. Strip mining—stripping away the surface of the earth to get at the goodies beneath—was the rule rather than the exception.

Reclaiming strip-mined land requires patience.

So The Wilds butterfly habitat, which has been carefully rehabilitated over a period of years, holds special significance. If the variety of butterfly species found in this small area of semi-restored prairie is any indication, the devastation can be ameliorated. Time will be the only true healer of this devastated land, but human efforts can pay off to a considerable degree. Common species seen here now include monarchs, cabbage whites, clouded and cloudless sulfurs, pearl crescents, and Delaware skippers. Also seen are gray hairstreaks; pipevine, black, and spicebush swallowtails; question marks; and great spangled fritillaries.

It takes a long time, thousands of years, to replace the carbon on the land surface that strip mining removes. Without carbon, the butterflies disappear along with other insects and the rest of animal life.

No carbon no plants.

No plants no animals.

No us.

The math is that simple.

Meanwhile, elsewhere along the butterfly highway, observers continued to report with great excitement. Monarch numbers continued to look spectacular. On the Kansas-Colorado border, five hundred were seen gathered on a tree as early as mid-September. "I've never seen so many," reported someone in Claremore, Oklahoma, on October 5. Around the same time in Ropesville, Texas, they were seen arriving over the course of an entire week. "Absolutely beautiful," reported another observer in a nearby town.

By mid-October in Hobbs, New Mexico, on the border with Texas, thousands were reported roosting in a cemetery. Hundreds were seen in a tree in Abilene, Texas, where, the observer reported, they had roosted repeatedly over the years. The butterflies were beginning to "funnel," that is, to gather into larger and larger groups as they neared Mexico. By the time they crossed the Mexican border, they had become a rushing stream.

In Tulsa, where I was soon headed, the local newspaper reported "hundreds of thousands" moving through. "They're making a comeback!" announced the Conservation Coalition of Oklahoma. In Bixby, just south of Tulsa, on October 6 a citizen scientist reported to Journey North: "Monarchs east to west, north to south, as far out and up as I could see with 10X binoculars. . . . a steady flow going south on the wind, with 20 to 40 in view at any one time." Along the western and eastern flyways, a different storyline played out. David James in Washington State was reporting near-disastrous numbers of overwintering populations up and down the California coastline. "Nobody knows for sure why our numbers are down so low," he told me.

The offspring of the butterflies that had managed to survive the previous winter in California weren't "that great," he continued. "We always go down at the end of May on Memorial Day to a certain spot on the California-Oregon border. We saw the lowest number that we've seen in the last five years. Something wasn't quite right."

And at Crab Creek, where we had met in that horrendous heat the previous year, James found nary a monarch throughout the whole of the summer of 2018: "Not a single one. They just didn't arrive. They got to Washington, but only along the state border. In the whole of central Washington, there wasn't a reliable sighting."

I asked if this might have had something to do with the fires that had burned once again throughout the region.

"This was in June, before the fires," he said. "It just didn't happen."

There was, however, good news east of the Rockies. I spoke with butterfly observer Gayle Steffy in Pennsylvania, who explained that she was a "monarchs for life" person. When she was thirteen, she found a monarch caterpillar in a field near her home, which she raised and released.

When she was fourteen, she and her brother found a book about monarchs in their local library. In the back of the book, it gave Fred Urquhart's name and address, in case readers wanted to participate in his tagging program. She wrote, but got a letter back from him announcing that tagging was "over."

So she created her own tags and her own tagging program. On the tag, she provided a post office box so that she could be contacted by anyone who found her tags.

Eventually she did get a letter—from Mexico. And so now, forty years later, she is still at it. Now using tags from Monarch Watch, an influential nonprofit I was soon to learn a lot more about, she keeps on counting monarchs and tagging monarchs and planting milkweed and all kinds of nectar plants, every year.

Steffy recently published a paper with thirty years of monarch data. "When I started crunching the numbers, I found out that early migrants are more successful, early migrants tend to be larger than later migrants, and that early migrants tend to be male rather than female."

This was a curious fact, I suggested.

"Females on average are smaller than males," she answered. "That may be why."

Steffy said that her observations this year suggested that monarch numbers east of the mountains seemed to be quite high, but that the insects were not ending up in Cape May, New Jersey, well known as a location where many migrating monarchs stop to enjoy the flowers for a few days before continuing their southern journey. Instead, she said, the winds were so favorable that they were heading down the western side of Chesapeake Bay.

Why, I asked, were numbers so high in her area this summer?

"The rain," she suggested. It wasn't just that the overwhelming amount of rain was good for the plants and their flowers. It was also that fields that were routinely mowed several times a summer had been mowed less often.

"The rain was a blessing and a curse. I have one field that got flooded and everything got swept away. But I think it was also a blessing, because roadsides and many fields could not be mowed at all. Everything was too wet." That was certainly one factor I hadn't yet heard of.

In recent years, Steffy had suffered a tragedy. The two sites along the Susquehanna—a power plant and a road-building site—that she had monitored for decades were destroyed by herbicide spraying.

She got a grant from the corporation she works for to plant, along with others on her sustainability team, 2,000 pollinator plants. She's also filling her own yard with butterfly-friendly plants.

"I'm creating my own habitat," she said.

When I arrived in Tulsa in late October, this slowly revitalizing city, devastated decades earlier by an oil market crash, remained awash in monarch butterflies. It was nearing the end of the migration, but the insects were still coming. On an afternoon stroll through Gathering Place, a privately funded, recently opened riverside park, at least thirty monarchs flitted from nectar source to nectar source, fueling up for the final push to Mexico.

Nearby, monarchs and plenty of other butterflies enjoyed the abundance of still-flowering plantings on the grounds of the world-famous Gilcrease Museum. Tulsa has done a terrific job of providing nectaring areas for butterflies.

Outside city limits, butterflies were fewer, since there were fewer planted butterfly gardens, but they were still around. An hour north, in the Osage Hills at the Nature Conservancy's 39,650-acre Tallgrass Prairie Preserve, most of the foliage had already senesced. The massive preserve, home to lots of bison and four different species of prairie grasses, including big bluestem that grows in some places to heights of nine feet, hosts roughly 100 different species of butterflies. At least nine species of those delicate little blues so loved by Nabokov—marine blue, western pygmy blue, eastern tailed blue, spring azure, summer azure, silvery blue, Jack's silvery blue, Reakirt's blue, Texas blue—thrive here throughout the entire season from spring to late fall.

Even though to my eyes most blooms were gone, the final waves of migrating monarchs searched there for food. They needed to make it to the Texas border within a day or so, the monarch scientist Chip Taylor told me, or the impending cold would prematurely end their lives. At well over 700 miles, the route was already depleted of nectariferous blooms. The future for these last stragglers did not look promising.

I was there to attend an event unique to Oklahoma: a meeting of the Tribal Alliance for Pollinators. Begun in 2014, the Alliance now includes seven of the thirty-nine tribes living in the state: Chickasaw, Seminole, Citizen Potawatomi, Muscogee (Creek), Osage, Eastern Shawnee, and Miami Nations. The purpose of the Alliance is to provide funding, training, and support to members of those tribes who wish to restore native plant habitats on tribal lands.

The three-day gathering began with a walk. Under overcast skies that foretold the soon-to-arrive winter, Taylor, eighty-one, and Andrew

Gourd, thirty-one, a member of the Seneca-Cayuga Nation, led a group of us through a field of several acres. Located up in the northwestern corner of the state, in a region known as the "green country," the countryside had already turned brown for the winter. But these browns were still interesting. For Taylor and Gourd, these few acres composed a gold mine of glorious depth, a labyrinth of epochal proportion.

I myself felt as though I'd reached the Promised Land. After two years of visiting butterfly habitat that had been vastly improved after much abuse, after looking at land that had been brought back from the brink—like the strip-mined lands of Ohio or the revitalized ranchlands of the Willamette Valley or the much-abused pine barrens near Albany, New York—I was finally getting to see land that looked somewhat like it had looked before European colonization.

This land was the real McCoy—the kind of land that, the soil microbiologist Nicola Lorenz explains, required several thousand years to evolve. As far as anyone could tell, this land was pretty much the way it had been for decades upon decades and centuries upon centuries.

"Never turned nor tilled," Gourd told Taylor.

"As you drive through Oklahoma, think about the land," Taylor had told the gathering before the walk. "Think about what the landscape must have looked like then [before homesteading], because it doesn't look like that now."

Taylor did not begin his career as a monarch researcher. Instead, after studying bees he shifted gears. With Lincoln Brower he began researching monarch migration issues. Today, as a university professor emeritus in Kansas, he is the founder and head of Monarch Watch, the nonprofit that supplies volunteers east of the Rockies with tags for butterflies. Begun in 1992, Taylor's program now encourages the planting of bloom-rich "Monarch Waystations" along migratory flyways and supplies 400,000 tags for volunteers (Gayle Steffy among them) yearly. When butterflies with tags are found in Mexico, Monarch Watch pays $5 for each recovered tag, then checks the data to determine where across North America the insects were tagged.

The program, he said, had revealed a wealth of data about monarch behavior that would not have otherwise been discovered. "This data is just so important. It answers questions about the size of successful migrants. It answers questions about origins. It answers questions about mortality. It answers questions about orientation. It answers questions about conservation. It just answers a tremendous number of questions."

To date, Taylor said, Monarch Watch has assembled something like 1.6 million bits of data. So much to analyze. So little time. Taylor has a phenomenally busy speaking schedule. From the end of October to the end of the year, he had five more events planned, along with the Thanksgiving and Christmas hiatus. He had just returned from the national capital, where he had been given a spectacular award—a monarch encapsulated in an exquisite glass weight—at the North American Pollinator Protection Campaign International Conference.

As he showed it to me, he looked pleased, but also tired.

I asked later why he continued to work so hard.

"Why not work until they carry you out?" he asked in return. "What are we here on this planet for? Just for our own gratification? Some people get gratification out of trying to make the world a better place. That's where I come from. I 'm going to try as long as I can. I enjoy what I do."

In many areas of the state, Taylor told the group, when he looked at a piece of land he was hard-pressed to find ten or fifteen plant species on any one parcel. In comparison, healthy prairie parcels may have well over 100 species. Those plants that he did find had extremely shallow roots, roots that did not penetrate the ground deeply. This was a problem during Oklahoma's sometimes years-long droughts.

On this Seneca-Cayuga land that we walked, we easily discovered at least forty species within the first minutes. Many of those plants had roots that delved deep into the soil, as much as six or ten or twenty feet below the surface. This made them drought resistant, since they could tap into moisture reservoirs that modern shallow-rooted grasses cannot access.

"Look at the diversity across this field," Taylor continued, "without a break in the diversity. That's probably what it looked like all across Oklahoma at one time."

Or all across the tall grass prairies that covered central North America. Plowing up that thickly rooted soil required teams of as many as thirty hefty, muscular plow horses. And once the sod was separated from the ground, it was thick enough to be used to build "sod houses," which kept families warm in winter and cool in summer because the roots provided so much insulation from the elements. You wouldn't be able to do that with modern grass sod, flimsy and shallow-rooted.

Taylor reached out to get some seeds from a grizzly-looking thing commonly called a "rattlesnake master." Standing five to six feet tall, with prickly leaves like yucca and thorns that reminded me of thistle plants, rattlesnake master is inescapably alluring to pollinators. Taylor pointed out two species of square-stemmed mints, compass plants, "six or ten species that you don't see unless the field has been there for a long, long time. Seeds like these don't get around easily. They're not brought by floods, not carried by birds . . ."

"This land really is a sacred place," Gourd said. "It is on a piece of property owned by Seneca-Cayuga tribal members and is truly pristine. It's on a hilltop that managed to avoid logging. It probably looks just like it did when our ancestors got here in 1831."

The federal government created a land allotment program for Indian families in 1887. Whether they wanted to own land or not, each tribal member was assigned a parcel. This particular land, part of the Cowskin Prairie of the Ozark Mountains foothills, was given to a member of the Whitetree family, Gourd said. The tribal member wanted to be left alone and farmed only twenty of the eighty acres received.

The remainder of the acreage, including this prairie remnant, simply fell by the wayside. The soil was rocky and would have required terracing, because it was on a slope. Forest was slowly encroaching and would have required logging. From the point of view of modern capitalism, it was worthless. No one wanted to own it. No one wanted to work it. All this makes this land priceless as an irreplaceable model for people living in the twenty-first century.

"If you walk a section line east from there," Gourd said, "the land would have been grazed and tilled. You wouldn't have had all that beauty. If you go south, you can see corn, a tree farm, whatever else that can take advantage of rich prairie soil. There wasn't any kind of restoration or holding plan in place. It was just the luck of the draw that it managed to survive."

I asked Gourd if he thought it would survive into the future.

We agreed that land like this was a national treasure, more important than gold or silver.

In California, by the end of the migration season, it looked like it was already too late to speak for the land. Or for the butterflies. Tens of thousands of acres of fires raged across the state, making the blazes of the year before look like mere campfires. Those fires took both the homes of poor people in the mountains, in towns with ironic names like Paradise, and the homes of the rich and beautiful and powerful in places like Malibu. By the conflagration's end, nearly a hundred souls were known to have perished. Another thousand or so were missing. In Butte County, north of Sacramento, well over 150,000 acres burned. Nearly 20,000 buildings were destroyed. The fires spread quickly due to the chaotic climate of recent years. There had been an inordinate dry spell on the West Coast, while east of the Rockies we experienced the wettest fall ever recorded.

What happened to California's migrating butterflies was anyone's guess. All we know is that they never got to their destinations. After performing their Thanksgiving 2018 count, the Xerces Society of Oregon reported that the entire western population had declined by 87 percent. Overall, they declared the western population to be "critically low."

At Pismo Beach, where the docent two years earlier had instructed the children in the story of the Butterfly Hotel, and where Kingston Leong and I had first met, and where tens of thousands of butterflies had overwintered for as long as anyone could remember, only 800 butterflies showed up. At the Morro Bay Golf Course, only 2,587 butterflies were found.

"Reasons for the drastic decline are not really known," Leong explained. He suspected that the large fires in Washington State and Canada over the summer, coupled with the disastrous California fires during the fall migration, were an important cause. One study Leong had done showed that the butterflies are "very sensitive to smoke."

"Hence, the fires that occurred in the fall may influence their migration to their coastal winter sites." Scientists are still studying this question.

On the Texas/Mexico border, the migratory flights had become hung up. Torrents of rain and unfavorable winds rushing across the Texas plains had caused multitudes of butterflies to huddle together, trying to keep warm.

"They won't go hungry, but first they're going to have to settle down until things warm back up," wrote Journey North contributor Dale Clark of Glenn Heights, Texas, on October 14, 2018. "I'll be interested to see what happens once this horrible cold spell and rain finally ends."

Then, just before Thanksgiving, the reports began to flow in: "We saw an average of about 10 per minute," one Mexican commentator wrote from Querétaro. They were in Valle de Bravo, and elsewhere throughout the mountains. The first arrivals had been reported as early as November 7, and by the time of the American Thanksgiving, monarch numbers looked healthy and strong.

Taylor had predicted this outcome six months earlier, at the start of the northern migration season.

Epilogue

IN THE MOUNTAINS OF MEXICO

Look about you at the little things that run the earth.

E. O. Wilson

At 10:00 a.m., right on time, a cloudburst of monarchs descending from their roosts met me head-on. A tangled palette of glittering color. Fantastical but nonetheless real. River of monarchs above a mountain stream, heading downward out of the forest and into the sun. All around me. I am one of them.

Gobsmacked again.

I consider myself jaded. Nearing my seventieth year, I'm a been-there-done-that oldster with a lot of world adventure under my belt. In my twenties I rode horses in Africa, galloping up and down sand dunes in the Sahara Desert with American Marines. In my thirties, I paddled the Okefenokee Swamp for a week, beginning a life as a travel writer. I have ridden elephants (not my favorite) and camels (*most definitely* not my favorite) and biked along many of America's best bike paths and walked the fossil-filled hills of Provence and walked among the wild horses of Mongolia.

When I went to the mountains of Mexico as the final step in researching this book, I did not expect an otherworldly experience. I'd seen a lot of butterflies over the past two years, including a great many monarchs. I did not expect to be yet again spellbound by sunlight and color. But I was.

The experience was poignant. I wondered how Maria Sibylla Merian would have felt. In the bright mountain sunlight, the same bedazzlement that

overwhelmed me when I saw that first Turner and when I saw those first butterfly boxes at Yale University took hold of me once more. As I trudged slowly
up the steep high-altitude climb to the mountain peaks of the El Rosario entry
point into the Monarch Butterfly Biosphere Reserve, the clouds shifted and
the forest was filled with energy and sunlight.

The insects halted their descent and came to rest on the shrubs on both
sides of the trail, spreading their wings to soak up the light. Standing in the
warmth of the sun in the center of all that stained-glass-window color circling
my head, I easily understood why local people celebrated the arrival of these
insects each fall.

The migration of the monarchs from points as far north as Canada all
the way south to these particular mountaintops is a world phenomenon that
belongs to everyone on the planet. It's a source of global joy, like the migration
of the wildebeest on the Serengeti Plain or the migration of gray whales off
the west coast of North America.

They are all following the sun, just as we would if we could.

And yet, one by one, these migrations are disappearing. Passenger pigeon
migration: Gone. North American bison migration: Vanished. Reindeer migration: Seriously decimated.

In the midst of all this, Amelia and her monarch butterfly give us hope.
Now, drunk with sunlight and monarch colors, I felt that hope once again,
walking up the mountain path. Well-groomed though it was, the high-
altitude walkway was steep. I live by the ocean and prefer my atmosphere
oxygen-rich.

I stopped frequently to catch my breath and to watch the multitudes—
the multitudes of descending insects, but also the multitudes of ascending
human beings. They reminded me of the pilgrims I'd seen walking along
Spain's Camino de Santiago, or the pilgrims who were even then filling the
roads leading into Mexico City to the Basilica of Guadalupe.

Most of the people who passed me by on the mountain trail as I stood
and watched were not American tourists, as I had expected. They were Mexican. The family I remember most clearly: supported by three or four younger
people, an elderly man who could barely move slowly inched his way up to the
peaks where still more butterflies rested on still more branches.

Obviously in a lot of pain and discomfort, he was determined to get to the top. With one arm around one younger man and another arm around a younger woman, he put one foot in front of the other, never giving up.

"Why?" I asked José Luis Paniagua, my world-class guide who had brought me from Mexico City for the day.

It had to do with family and ancestors, he explained. They all came as a family to see the butterflies and wanted to have that experience together. No matter how difficult this trek was for him, he would have wanted to be part of the family. They would never have left him behind.

Butterflies unite us across generations and across space and across time. They are elemental. A butterfly is an entire universe, right there in the palm of your hand. As toddlers we reach for them instinctively. As children, we chase them. As adults we study them and learn how essential they are to our entire world. As we age, we see their gorgeous colors as something to love in our waning years.

The monarch joins together the Mexican family with their elderly grandfather in the Mexican mountains with Amelia and her mother, Molly, in the Willamette Valley, joins together Andrew Gourd, thirty-one, of the Seneca-Cayuga people with Orley "Chip" Taylor, eighty-one, the scientist from Kansas who has chosen to spend his last years working to protect this insect.

Butterflies unite people around the world, but also across time, from the inestimably brave Maria Sibylla Merian to the inestimably pensive Charles Darwin to the scores of contemporary scientists who continue to unlock the secrets of the world's favorite insect. And yet there's so much more to learn.

"We managed to get a man on the moon before we discovered where the monarchs went," my butterflying friend Joe Dwelly pointed out one day at lunch.

Sadly, despite these centuries of hard work, butterfly numbers are diminishing. Indeed, scientists suspect that the entire group of living creatures that we call insects are suffering severe population depletions. There are banner years, to be sure. As I write this, butterfly monitors are celebrating the clouds of painted ladies that are showing up in the northern reaches of both the Eastern and the Western Hemispheres. But the numbers are stochastic, and the general trend shows a clear downward slope.

There are most likely a thousand—or a hundred thousand—reasons why. Complex, lush fields full of native nectar-filled plants have been turned into agrobusiness-dominated monocultures. Grass lawns cover vast acreages that were once resplendent with wildflowers. Pesticides are so prevalent that they now commonly pollute our drinking water and are a part of our body chemistry.

The climate chaos that I saw everywhere I went during my two years chasing butterflies is having an inestimable effect. Butterflies that are highly adapted to specific situations, like the delicate blue butterflies loved by Nabokov, have no chance against the roller-coaster climate we now live in.

But there are other, as yet not discovered, reasons why butterflies are disappearing. Research has revealed that monarch caterpillars that consume milkweed plants growing by some roadsides are *saltier* than caterpillars on milkweed plants growing by other roadsides. The difference: whether or not local road departments salt their roads in the winter. We are likely beginning a major period of evolutionary adjustment and a concomitant period of species extinctions.

It doesn't have to be this way. We already have proof of concept. Once scientists unlocked the secret lifestyles of the small blue butterflies, they were able to bring them back from the brink.

If we are determined, we can accomplish great things. But why bother?

Us oldsters can remember a world of rich natural beauty, a world in which each month of the year brought new smells, new sounds, new sights, new promises of absolutely essential connections between humans and their natural surroundings.

That world is fast disappearing. But it's not gone yet. We can get it back. When a five-year-old girl releases a butterfly into the air, and when that butterfly is seen by other people as it flies to its winter destination, that, to my mind, is the real butterfly effect: the joining together of countless people of many different nations, across generations, in a united effort to protect at least one small joyful piece of the natural world to which we belong.

Acknowledgments

An overview of several hundred years of butterfly history depends entirely on the kindness of strangers—scientists, citizen scientists, historians, authors, and just plain lovers of Lepidoptera. When I began this work, I wasn't sure how interested others would be in discussing their work. You never know for sure.

I was overwhelmed with kindness. People met me for a day, sometimes for several days. Scientists spent hours explaining their work, then, if I needed more help, agreed to talk again. The world of science has changed considerably since I began working as a science journalist forty years ago, when many "name" researchers did not want to take the time to communicate to me and to the public.

This generosity turned out to be particularly true in researching this book. When it comes to Lepidoptera, from scientists and enthusiasts alike, the enthusiasm for communication was palpable. Among the many who welcomed me were Amelia Jebousek and her mother, Molly, who explained the ecology of the Willamette Valley in thorough detail as we spent a roasting-hot day driving around farming and wetlands regions. David James kindly spoke to me on the phone many times as I followed his work, and visited with me on two separate occasions. Kingston Leong honored me by showing me his many monarch revitalization projects. Thanks also to Adriana Briscoe, Josh Heptig, Anurag Agrawal, Matthew Lehnert, Jennifer Zaspel, Konstantin Kornev, Peter Adler, Warren and Laurie Halsey, Michael Engel, Neil Gifford, Herbert Meyer, Ricardo Pérez–de la Fuente, Conrad Labandeira, Susan Butts, Chris Norris, Jim Barkley, Gwen Antell, Jessica Griffiths, Mia Monroe, Patrick Guerra, Steve Reppert, Cathy Fletcher, Hugh Dingle, Micah Freedman, Gerard Talavera, Nipam Patel, Richard Prum, Radislav Potyrailo, Lincoln

ACKNOWLEDGMENTS

Brower, Karen Oberhauser, Elizabeth Howard, Gayle Steffy, Andrew Gourd, Chip Taylor, Joe Dwelly, Kate Hunter, Linda Kappen, Steve Malcolm, Jeff Glassberg, Cheryl Schultz, and the many butterfly fans who willingly talked about their passion wherever I went.

Thanks to all the folks at Simon & Schuster: Rebecca Strobel, Molly Gregory, Karyn Marcus, and Kayley Hoffman; to Math Monahan for the exquisite cover design; to agent Michelle Tessler; to Annie Gottlieb, an exceptional copyeditor and friend; and to SallyAnne McCartin, whose decades in the book business make her advice invaluable.

Many thanks to my husband, Greg Auger, for his helpful photographs.

One more great big thanks to Denise McEvoy, whose kindness in caring about all of Earth's living creatures has meant so much to so many.

Notes

vii "Nature has a perverse preference": Michael S. Engel, *Innumerable Insects: The Story of the Most Diverse and Myriad Animals on Earth* (New York: Sterling, 2018), xiii.

Introduction

11 "Color is a power": Wassily Kandinsky, *Concerning the Spiritual in Art* (Munich, 1911).

Part I: Past

One. The Gateway Drug

3 "familiar with the speckles and dappling": Richard Fortey, *Dry Storeroom No. 1: The Secret Life of the Natural History Museum* (New York: Alfred A. Knopf, 2008), 55.

3 Herman Strecker was: Information on Herman Strecker is plentiful, but for the most in-depth discussion of his psyche, read William R. Leach, *Butterfly People: An American Encounter with the Beauty of the World* (New York: Pantheon, 2013).

4 He was "omnivorous": Leach, *Butterfly People*, 61.

4 "My soul pines": Leach, 61.

4 "Why did God implant": Leach, 199.

5 "a *duty* to inventory": Fortey, *Dry Storeroom No. 1*, 43.

5 "Collecting was a Victorian passion": Jim Endersby, *Imperial Nature: Joseph Hooker and the Practices of Victorian Science* (Chicago: University of Chicago Press, 2008), 54.

5 "a butterfly good-time": Walt Whitman, *Specimen Days and Collect*

(1883; repr. New York: Dover Publications, 1995), 121; quoted in Leach, *Butterfly People*, xviii*n*9.

7 "How little we know": Christopher Kemp, *The Lost Species: Great Expeditions in the Collections of Natural History Museums* (Chicago: University of Chicago Press, 2017), xv.

7 "Evolution begets diversity": David Grimaldi and Michael S. Engel, *Evolution of the Insects* (New York: Cambridge University Press, 2005), 1.

8 "By most measures": Grimaldi and Engel, *Evolution of the Insects*, 1.

8 "Without a doubt": Grimaldi and Engel, 4.

14 The idea was so horrifying: Michael Leapman's *The Ingenious Mr. Fairchild: The Forgotten Father of the Flower Garden* (New York: St. Martin's Press, 2001) is a charming book, originally published in Britain, that discusses the horror and controversy surrounding the unspeakable discovery that flowers have male and female parts.

Two. Down the Rabbit Hole

15 "as simple as a butterfly": Destin Sandlin, Deep Dive Series #3: "Butterflies," *Smarter Every Day*, educational video channel, http://www.smartereveryday.com/videos.

15 "writing a letter to his best buddy": Darwin wrote endlessly, but this is one of his most famous letters, in part because of his question about the insect, and in part because it so endearingly details his personal life. Almost all Darwin's letters are now available online. You can read this letter in its entirely here: https://www.darwinproject.ac.uk/letter/DCP-LETT-3411.xml.

18 Four-year-old Matthew Lehnert: Lehnert is currently at Kent State University at Stark, in Canton, Ohio, where he divides his time between teaching and authoring numerous research papers with titles like "Proboscis Morphology Suggests Reduced Feeding Abilities of Hybrid *Limenitis* Butterflies (Lepidoptera: Nymphalidae)." That one, from *Biological Journal of the Linnaeus Society* 125, no. 3 (2018): 535–46, can be found here: https://academic.oup.com/biolinnean/article-abstract/125/3/535/5102370.

21 Entomologist Jennifer Zaspel knows: Jennifer Zaspel et al., "Genetic Characterization and Geographic Distribution of the Fruit-Piercing and

Skin-Piercing Moth *Calyptra thalictri* Borkhausen (Lepidoptera: Erebidae)," *Journal of Parasitology* 100, no. 5 (2014): 583–91.

22 "Humans experience pain": Harald W. Krenn, "Feeding Mechanisms of Adult Lepidoptera: Structure, Function, and Evolution of the Mouthparts," *Annual Review of Entomology* 55 (2010): 307–27, https://www.ncbi.nlm.nih.gov/pmc/articles/PMC4040413/.

23 Peter Adler and Konstantin Kornev began: Adler and Kornev are both currently at Clemson University and continue to publish research about insect proboscises.

23 wanderings of two little girls: Konstantin Kornev, personal communication.

25 "to see if there are gene products": Jennifer Zaspel, personal communication.

26 "short and stub-like and fleshy": Matthew Lehnert, personal communication.

Three. The Number One Butterfly

27 "One would hardly anticipate": Samuel Hubbard Scudder, *Frail Children of the Air: Excursions into the World of Butterflies* (Boston and New York: Houghton, Mifflin, 1897), 268.

27 34 million years ago: The visitor center at the famous Florissant Fossil Beds National Monument (https://www.nps.gov/flfo/index.htm) is a fount of information.

27 this natural Notre Dame: Herbert W. Meyer's *The Fossils of Florissant* (Washington, DC: Smithsonian Books, 2003) provides an excellent introduction to the entire ecosystem of the region from 34 million years ago.

30 Theodore Mead: Theodore Luqueer Mead deserves much more attention in a book about butterflies than I was, sadly, able to give him. He loved both butterflies and plants. The Mead Botanical Garden in Winter Park, Florida, is named after him. Some credit him with "discovering" the Florissant fossil beds, but of course that isn't accurate. Many people already knew about them. What he did do was send a large sample to Harvard's Samuel Scudder, who then made the discoveries famous worldwide.

31 "sacred ground for American paleontology": Kirk Johnson and Ray Troll

(illustrator), *Cruisin' the Fossil Freeway: An Epoch Tale of a Scientist and an Artist on the Ultimate 5,000-Mile Paleo Road Trip* (Golden, CO: Fulcrum, 2007), 180.

31 Walt Disney came by: Meyer, *Fossils of Florissant*, 15–17.

32 "boxes upon boxes full": "A Celebration of Charlotte Hill's 160th Birthday," *Friends of the Florissant Fossil Beds Newsletter* 2009, no. 1 (April 2009): 1.

32 one of Charlotte's biggest fans: Herbert Meyer, personal communication. Meyer has worked for years to correct the omission of Hill from the scientific record, believing that she did not get credit for her work because she was a woman and worked without formal credentials. He has written about her copiously, including in Estella B. Leopold and Herbert W. Meyer, *Saved in Time: The Fight to Establish Florissant Fossil Beds National Monument, Colorado* (Albuquerque: University of New Mexico Press, 2012).

32 "In ancient times": William A. Weber, *The American Cockerell: A Naturalist's Life, 1866–1948* (Boulder: University Press of Colorado, 2000), 62.

33 "so perfect as to allow a description of the scales": Samuel H. Scudder, "Art. XXIV.—An Account of Some Insects of Unusual Interest from the Tertiary Rocks of Colorado and Wyoming," in *Bulletin of the United States Geological and Geographical Survey of the Territories*, ed. F. V. Hayden, vol. 4, no. 2 (Washington, DC: Government Printing Office, 1878): 519.

33 "a picture of this fossil butterfly": Liz Brosius, "In Pursuit of *Prodryas persephone*: Frank Carpenter and Fossil Insects," *Psyche: A Journal of Entomology* 101, nos. 1–2 (January 1994): 120.

34 tried to sell the fossil: Herbert Meyer, personal communication.

36 "insectan Pompeii": David Grimaldi and Michael S. Engel, *Evolution of the Insects* (New York: Cambridge University Press, 2005), 87.

36 Yet the site was almost lost: Estella Leopold and Herbert Meyer's short book *Saved in Time* is an in-depth discussion of how this invaluable property was set aside for science and for the public, despite its obvious value to real estate speculators. I love books that detail the histories of pieces of land. It's so important for us to know from whence came the public lands that we sometimes take for granted.

36   "a page of earth history": Leopold and Meyer, xxiv, 45.

36   "wrapping fish in the Dead Sea Scrolls": Leopold and Meyer, 76.

36   "Rosetta Stone for grinding corn": Leopold and Meyer, xxvi.

37   Green River fossils: My favorite general book on this remarkable fossil-iferous region is Lance Grande, *The Lost World of Fossil Lake: Snapshots from Deep Time* (Chicago: University of Chicago Press, 2013).

Four. Flash and Dazzle

45   "The wings of a butterfly": G. Evelyn Hutchinson, quoted in Naomi E. Pierce, "Peeling the Onion: Symbioses between Ants and Blue Butter-flies," in *Model Systems in Behavioral Ecology: Integrating Conceptual, Theoretical, and Empirical Approaches*, ed. Lee Alan Dugatkin (Prince-ton, NJ: Princeton University Press, 2001), 42.

45   Maria Sibylla Merian would become famous: In recent years this long-in-visible genius, mother, and housewife has become much better known in the English-speaking world, but since most of her work has not been trans-lated into English, she remains somewhat mysterious to us. That began to change in the 1990s, when historian Natalie Zemon Davis included Merian in *Women on the Margins: Three Seventeenth-Century Lives* (Cambridge, MA: Harvard University Press, 1995). After that, biologist Kay Etheridge has led the way in touting Merian as the founder of ecology, in papers such as "Maria Sibylla Merian: The First Ecologist?," in *Women and Science: 17th Century to Present: Pioneers*, ed. Donna Spalding Andréolle and Véronique Molinari, 35–54 (Newcastle upon Tyne, UK: Cambridge Scholars, 2011), http://public.gettysburg.edu/~ketherid/merian%201st%20ecologist.pdf; and "Maria Sibylla Merian and the Metamorphosis of Natural History," *Endeavour* 35, no. 1 (March 2011): 16–22, https://www.sciencedirect.com/science /article/pii/S0160932710000700. In May 2014 the first conference on Me-rian was held, and Etheridge became a board member. She is currently hav-ing some of Merian's work translated into English.

47   "If they did connect a larva": Michael S. Engel, personal communica-tion.

48   "a remarkably complex challenge": For a fascinating and well-written discussion of how difficult it was for people to make the seemingly ob-vious connection between caterpillar and butterfly via chrysalis, read Matthew Cobb's *Generation: The Seventeenth-Century Scientists Who*

Unraveled the Secrets of Sex, Life, and Growth (New York: Bloomsbury, 2006); quote, 134.

49 "a woman could give birth to a rabbit": Cobb, *Generation*, 222.

50 Yet Merian got away with it: The best set of essays on Merian currently available in English can be found in the remarkable reproduction of Merian's *Metamorphosis Insectorum Surinamensium*, made available in 2016 by Lannoo Publishers, a Dutch venture. This modern version can be ordered either through your local bookstore or online. It's the precise size of Merian's original book and contains excellent reproductions of her artwork. It also contains her original essays on each individual artwork, both in Dutch and translated into English. At the beginning of the entire reproduced volume are essays by top Merian scholars, including Kay Etheridge. In the back of the reproduced book is a list of remaining originals (there are not many left, most having been disassembled and sold piecemeal) and where they may be found.

51 "one of the most extraordinary figures in the history of science": Gauvin Alexander Bailey, "Books Essay: Naturalist and Artist Maria Sibylla Merian Was a Woman in a Man's World," *The Art Newspaper*, April 1, 2018, https://www.theartnewspaper.com/review/bugs-and-flowers-art-and -science.

51 Attenborough featured her work: David Attenborough et al., *Amazing Rare Things: The Art of Natural History in the Age of Discovery* (2007; repr. New Haven, CT: Yale University Press, 2015).

51 "curious, willful, self-concealing, versatile": Zemon-Davis, *Women on the Margins*, 141.

56 "her beloved insects": Entomologist Michael S. Engel's well-written and highly accessible *Innumerable Insects: The Story of the Most Diverse and Myriad Animals on Earth* (New York: Sterling, 2018) pays great homage to the foundation laid by Merian for the field of entomology. This particular quote can be found on p. 96, along with a sample of her artwork.

56 "a significant tributary": Etheridge, "Maria Sibylla Merian: The First Ecologist?"

58 butterfly lover Richard Prum: This story can also be found in Prum's work *The Evolution of Beauty: How Darwin's Forgotten Theory of Mate Choice Shapes the Animal World—and Us* (New York: Doubleday, 2017).

60 they have defense value: This theory was put forth in an essay by famed

entomologist Thomas Eisner in 2007, written shortly before his death: "Scales: On the Wings of Butterflies and Moths," *Virginia Quarterly Review* 82, no 2 (Spring 2006). It can be accessed here: https://www .vqronline.org/vqr-portfolio/scales-wings-butterflies-and-moths.

62 As a condition of acceptance: Nipam Patel, personal communication.

63 like a pliable plastic bag: This clever analogy was given to me by Yale's Richard Prum.

65 Gyroids, as Schoen imagined them: Alan H. Schoen, "Infinite Periodic Minimal Surfaces without Self-Intersections," NASA Technical Note D-5541 (Washington, DC: NASA, 1970).

65 a human-made 3-D structure: Zongsong Gan et al., "Biomimetic Gyroid Nanostructures Exceeding Their Natural Origins," *Science Advances* 2, no. 5 (2016): e1600084, https://advances.sciencemag.org/content/2/5 /e1600084.full.

66 butterflies can change color: Jim Shelton, "Butterflies Are Free to Change Colors in New Yale Research," *Yale News*, August 5, 2014, https://news.yale.edu/2014/08/05/butterflies-are-free-change-colors-new-yale-research.

Five. How Butterflies Saved Charles Darwin's Bacon

67 "It was snowing butterflies," "large and brilliant butterflies": These phrases come from *The Voyage of the Beagle* [*Journal of Researches*, 1839], which exists in many editions. This adventure travelogue, the first of Darwin's successful popular works, is meant for the lay reader and is highly accessible. To my mind, it's just as good as any of the adventure novels written during the same era, like Johann David Wyss's *Swiss Family Robinson* or Robert Louis Stevenson's *Treasure Island*.

69 Bates remained in South America: Henry Walter Bates wrote his own adventure travelogue, *The Naturalist on the River Amazons* (the "s" is correct), published in 1905. But despite his importance to science, no definitive biography of him has been written, sad to say. British author Anthony Crawforth wrote *The Butterfly Hunter: The Life of Henry Walter Bates* (Buckingham, UK: University of Buckingham Press, 2009), but this charming little book is as much about Crawforth's adventures in South America as it is about Bates. Sean B. Carroll's *Remarkable Creatures: Epic Adventures in the Search for the Origin of Species*

(Boston: Houghton Mifflin Harcourt, 2009) puts Bates's work in perspective for the lay reader, and of course Bates is discussed in great detail in most Darwin biographies. But he really does deserve his own biography.

70 "We cannot conceive that such marvellous perfection": T. V. Wollaston, "[Review of] *On the Origin of Species* [. . .]," *Annals and Magazine of Natural History* 5 (1860): 132–43, http://darwin-online.org.uk/content /frameset?itemID=A18&viewtype=text&pageseq=1.

70 deeply disturbed by Wollaston's hostility: Many (if not most) biographies of Darwin discuss his anxiety over the publication of his ideas. The Darwin scholar and biographer Janet Browne's second volume, *Charles Darwin: The Power of Place* (New York: Alfred A. Knopf, 2002), is an excellent source for this particular focus. I also love Adrian Desmond and James Moore's *Darwin: The Life of a Tormented Evolutionist* (reprint ed., New York: Norton, 1994).

71 "I have an immense number of facts": Bates to Darwin, March 28, 1861, Darwin Correspondence Project, https://www.darwinproject.ac.uk/letter /DCP-LETT-3104.xml.

71 "deceptive dress": Charles Darwin, "[Review of] 'Contributions to an Insect Fauna of the Amazon Valley,' by Henry Walter Bates [. . .]," *Natural History Review* 3 (April 1863): 219–24.

71 blending in with the badass majority: Many people have written about this key breakthrough. For a thorough yet accessible discussion, read chapter 4, "Life Imitates Life," in Sean Carroll's *Remarkable Creatures*.

72 "I think I have got a glimpse": Bates to Darwin, March 28, 1861.

72 "Contributions to an Insect Fauna": *Transactions of the Linnean Society* 23 (November 1862): 495, https://archive.org/details/contributionstoi00bate /page/502.

72 "overlooked in the ever-flowing rush": Darwin, review of Bates, "Contributions."

72 Albert Brydges Farn: A. B. Farn to Darwin, November 18, 1878, http:// www.darwinproject.ac.uk/letter/DCP-LETT-11747.xml.

73 "Butterflies were destined to become": Browne, *Charles Darwin: The Power of Place*, 226.

73   A third naturalist-wanderer: Fritz Müller is another researcher from the
      Darwinian time period who deserves a good English-language biogra-
      phy. For a brief description of Müller as a person and of the importance
      of his work, read Peter Forbes's *Dazzled and Deceived: Mimicry and
      Camouflage* (New Haven, CT: Yale University Press, 2009).

73   "insectivorous protection racket": Forbes, *Dazzled and Deceived*, 41.

74   Charles R. Brown and Mary Bomberger Brown: The Cliff Swallow Proj-
      ect, which bills itself as "one of the longest-running field studies of
      birds in North America," has been ongoing since the 1980s. Brown and
      Brown now head a team of researchers whose discoveries have informed
      us more deeply about how evolution works, and also about the lifestyles
      of these fascinating birds. Charles Darwin would have loved these two
      scientists. You can find out more about their work here: http://www
      .cliffswallow.org/.

74   the British peppered moth: The publishing history that surrounds this
      story goes on and on, and is a prime example of the kind of absurdity
      that can surround a simple science story when it becomes politicized.
      For a while, this story was widely accepted. But then anti-evolutionists
      began claiming that the science proving that the moth evolved was
      fraudulent. To set the record straight, the British researcher Michael
      Majerus began in 2001 a long-term study to test the evolutionary story.
      Majerus died in 2001 before publishing his paper. His colleagues saw
      the project through. It provided conclusive "evidence . . . implicat[ing]
      camouflage and bird predation" as the mechanism of natural selection
      for color change in the moths. It can be found online here: https://
      royalsocietypublishing.org/doi/full/10.1098/rsbl.2011.1136.

74   "incredibly easy to evolve these new colors": Rae Ellen Bichell, "But-
      terfly Shifts from Shabby to Chic with a Tweak of the Scales," NPR,
      August 7, 2014, https://www.npr.org/2014/08/07/338146490/butterfly-
      shifts-from-shabby-to-chic-with-a-tweak-of-the-scales.

76   "every day a splendid box of butterflies": Bates to his brother, quoted in
      Crawforth, *Butterfly Hunter*, 93.

Six. Amelia's Butterfly

79 "these are flowers that fly": Robert Frost, "Blue-Butterfly Day," from *New Hampshire* (New York: Henry Holt, 1923).

79 "Look about you at the little things": Edward O. Wilson, *Half-Earth: Our Planet's Fight for Life* (New York: Liveright / Norton, 2016), 111.

83 the popular monarch story: Fred A. Urquhart, "Found at Last: The Monarch's Winter Home," *National Geographic* 150 (August 1976): 160–73, http://www.ncrcd.org/files/4514/1150/3938/Monarch_Butterflies_Found _at_Last_the_Monarchs_Winter_Home_-_article.pdf.

85 After a journey of only nineteen days: The travels of Amelia's monarch were well-documented on the internet. Here are only a few of the stories: https://ucanr.edu/blogs/blogcore/postdetail.cfm?postnum=27559. https://news.wsu.edu/2018/06/25/monarch-butterfly-migration/.

85 David James maintains an active Facebook page, Monarch Butterflies in the Pacific Northwest, which keeps readers up to date on his tracking project's progress.

Seven. A Parasol of Monarchs

87 "a rain of golden sequins": Robert Michael Pyle, quoted in Sandra Blakeslee, "Butterfly Seen in New Light by Scientists," *New York Times*, November 28, 1986, A27.

90 Very recent research: The ecologist Andy Davis has released preliminary information that suggests that the butterflies may experience a stress response to extreme noise. Davis found elevated heart rates in caterpillars exposed to days of unrelenting stress, adding that some of his laboratory staff were bitten. https://www.upi.com/Science_News/2018/05/10 /Highway-noise-alters-monarch-butterflys-stress-response-could-affect-migration/5861525973774/.

91 Leong is famous: Kingston Leong has published a substantial number of research papers on the practical aspects of caring for monarch butterflies. A few of them can be found here: https://works.bepress .com/kleong/. And here: http://www.tws-west.org/westernwildlife/vol3 /Leong_WW_2016.pdf.

97   "From the mouth of the god": from Carlos Beutelspacher, *Las Mariposas entre los Antiguous Mexicanos* [Butterflies of Ancient Mexico], quoted in Karen S. Oberhauser, "Model Programs for Citizen Science, Education, and Conservation: An Overview," in *Monarchs in a Changing World: Biology and Conservation of an Iconic Butterfly*, ed. Karen S. Oberhauser, Kelly R. Nail, and Sonia Altizer (Ithaca, NY: Comstock / Cornell University Press, 2015), 2.

98   like a "cat drinking milk": Miriam Rothschild, quoted in Sharman Apt Russell, *An Obsession with Butterflies: Our Long Love Affair with a Singular Insect* (New York: Basic Books, 2003), 29.

98   drowning himself in the stuff: Anurag Agrawal, *Monarchs and Milkweed: A Migrating Butterfly, A Poisonous Plant, and Their Remarkable Story of Coevolution* (Princeton, NJ: Princeton University Press, 2017), 4.

99   "It just about knocked me over": Lincoln Brower, transcript of interview by Christopher Kohler, March 14, 1994, Oral History, University of Florida Digital Collections, 11, http://ufdc.ufl.edu/UF00006168/00001.

99   "One day, on tearing off some old bark": Darwin, *The Life and Letters of Charles Darwin, Including an Autobiographical Chapter*, ed. Francis Darwin, vol. 1 (1887; New York: D. Appleton, 1897; facsimile ed., High Ridge, MO: Elibron Classics / Adamant Media, 2005), 43.

99   found them similarly "vile": Nabokov, quoted in Robert H. Boyle, "An Absence of Wood Nymphs," *Sports Illustrated*, September 14, 1959, https://www.si.com/vault/1959/09/14/606166/an-absence-of-wood-nymphs.

99   "must eat milkweed": For the best layperson's information on the rigors of surviving on milkweed, read Agrawal's *Monarchs and Milkweed*.

102   "evolutionary back-and-forth": Michael S. Engel, personal communication.

102   "the ecological rise of flowering plants": Michael S. Engel, personal communication.

104   "nature's prime example of the male chauvinistic pig": Miriam Rothschild, "Hell's Angels," *Antenna: Bulletin of the Royal Entomological Society* 2, no. 2 (April 1978): 38–39.

105   She also knew a lot about fleas: Dame Miriam Rothschild is absolutely irresistible. Were she alive today (she died in 2005), I would have flown

to the ends of the earth to be able to meet her. Happily for all of us, she left behind some wonderful video interviews. One series of interviews, made for BBC TV in 1995 under the rubric "Seven Wonders of the World," can be found on the internet here:

> Part I https://www.youtube.com/watch?v=K2VaTmrsFLg
> Part II https://www.youtube.com/watch?v=fec8DClOhgo
> Part III https://www.youtube.com/watch?v=hRYcQmY5aTs

109 blue jays regurgitated monarch butterflies: Lincoln Pierson Brower, "Ecological Chemistry," *Scientific American* 220, no. 2 (February 1969), https://www.scientificamerican.com/magazine/sa/1969/02-01/.

109 enormously proud of this achievement: I spoke with Dr. Brower at great length on the phone only months before he died. At the time I had no idea he was that ill, although several of his colleagues had urged me to contact him quickly. We discussed his work in depth, and he advised me on a number of other people I should be sure to contact. Giving me his time at this point in his life was a great gift to me. I am continuously stunned by the devotion and caring of most of the scientists I am privileged to meet. For these people, science is not just a "job" or a "vocation." It is their reason for being. His *New York Times* obituary can be found here: https://www.nytimes.com/2018/07/24/obituaries/lincoln-brower-champion-of-the-monarch-butterfly-dies-at-86.html.

Nine. Scablands

111 "the most interesting insect in the world": Miriam Rothschild and Clive Farrell, *The Butterfly Gardener* (1983; reprint ed., New York: Penguin, 1985).

111 "shoved unceremoniously into Seattle's underbelly": Ellen Morris Bishop, *Living with Thunder: Exploring the Geologic Past, Present, and Future of the Pacific Northwest* (Corvallis: Oregon State University Press, 2014).

113 There were monarchs here?: David James usually holds public tagging sessions at Crab Creek on one or two weekends in August, depending on the timing of the migration south to California. One way to find out when he's holding these tagging sessions is to look at the website of the Washington Butterfly Association, an exceptionally vibrant nonprofit organization that strives to include members of the public as well as

professionals. You can find their website here: https://wabutterflyassoc
.org/home-page/.

Ten. On the Raindance Ranch

124 Wapato and camas, wild foods: wapato, http://www.confluenceproject.org
/blog/important-foods-wapato/; camas, http://www.confluenceproject.org
/blog/profound-role-of-camas-in-the-northwest-landscape/.

125 The Fender's blue is a homebody: The story of the recovery of this
delicate butterfly is remarkable. Its recovery plan appeared in the Federal Register of Tuesday, October 31, 2006, and can be accessed here:
https://www.fws.gov/policy/library/2006/06-8809.pdf.

126 Paul Severn, a modern butterfly fanatic: Many thanks to this lifelong
butterfly addict for the lengthy and enthusiastic phone conversation,
during which he discussed the details of his discovery of the remarkable
butterfly once believed to be extinct.

127 the Nature Conservancy bought a bog: David James, personal communication.

128 chain of refuges as stepping stones: Cheryl B. Schultz, "Restoring
Resources for an Endangered Butterfly," *Journal of Applied Ecology*
38 (2001): 1007–19, https://www.nceas.ucsb.edu/~schultz/MS_pdfs
/JAE%20Oct2001.pdf.

129 The butterfly in question: The story of the recovery of the large blue
butterfly in Britain is one of the most fascinating recovery stories I've
ever investigated. The long-term commitment of scientists and conservationists—and their refusal to quit when stumbling block after stumbling block arose—is a model for people who yearn to do something
about the planet's disappearing species. On September 19, 2018, *The
Guardian* ran a story saying that the insect had had "its best UK summer on record." Read the story here: https://www.theguardian.com
/environment/2018/sep/19/uk-large-blue-butterfly-best-summer-record.

130 It took thirty-five years: To get some sense of the immense effort
that went into unlocking the secrets of this butterfly, access this bro-
chure: https://ntlargeblue.files.wordpress.com/2010/06/large-blue-ceh
-leaflet0031.pdf.

130 "neurotic aristocrats": Matthew Oates, *In Pursuit of Butterflies: A Fifty-
Year Affair* (New York: Bloomsbury, 2015), 426.

131 Researcher Jeremy Thomas looked at all the red ants: J. A. Thomas et al., "Successful Conservation of a Threatened *Maculinea* Butterfly," *Science* 325, no. 5936 (July 2009): 80–83, https://science.sciencemag.org/content/325/5936/80.

133 "a Large Blue settled to bask": Oates, *In Pursuit of Butterflies*, 352.

Eleven. A Sense of Mystical Wonder

135 "[T]he highest enjoyment of timelessness": Vladimir Nabokov, *Speak, Memory* (rev. & expanded ed., 1967; Everyman's Library ed., New York: Alfred A. Knopf, 1999), 106.

136 "[M]y desire for it was one of the most intense": Nabokov, *Speak, Memory*, 120.

136 "a certain spot in the forest": Nabokov, 75.

136 "In the green group": Nabokov, 35.

137 "godfather to an insect": From Nabokov's exquisite poem "On Discovering a Butterfly," https://genius.com/Vladimir-nabokov-a-discovery-annotated.

137 Albany Pine Bush Preserve: The best resource for the history of this remarkable conservation project can be found in Jeffrey K. Barnes, *Natural History of the Albany Pine Bush, Albany and Schenectady Counties, New York: Field Guide and Trail Map* (Albany: The New York State Education Department, 2003).

141 "quite something to imagine": Robert and Johanna Titus, *The Hudson Valley in the Ice Age: A Geological History and Tour* (Delmar, NY: Black Dome Press, 2012).

142 Nabokov had been spot on: Carl Zimmer, "Nonfiction: Nabokov Theory on Butterfly Evolution Is Vindicated," January 25, 2011, https://www.nytimes.com/2011/02/01/science/01butterfly.html.

Part III: Future

Twelve. The Social Butterfly

149 "Mia Monroe, a volunteer": personal communication.

150 "blurring day into night": William Leach, *Butterfly People: An American Encounter with the Beauty of the World* (New York: Pantheon, 2013), 167. Leach's description draws on firsthand accounts of swarms

collected by B. D. Walsh and C. V. Riley in "A Swarm of Butterflies," *The American Entomologist* 1, no. 1 (September 1868): 28–29.

150 "almost past belief," "the heavens became almost black": Eyewitness accounts quoted by Lincoln Brower in "Understanding and Misunderstanding the Migration of the Monarch Butterfly (Nymphalidae) in North America," *Journal of the Lepidopterists' Society* 49, no. 4 (1995): 304–85.

150 the Canadian biologist Fred Urquhart: Urquhart's amazing story has been told numerous times, beginning with "Found at Last: The Monarch's Winter Home," *National Geographic* 150 (August 1976): 160–73, http://www.ncrcd.org/files/4514/1150/3938/Monarch_Butterflies_Found_at_Last_the_Monarchs_Winter_Home_-_article.pdf. In 1998, Urquhart and his wife, Norah, received the Order of Canada for their work, which was said to be "one of the greatest natural history discoveries of our time."

151 "We have found them—millions of monarchs!": Urquhart, "Found at Last."

152 We are all addicted to the sun: An excellent summary of the recent science, written for the layperson, is Russell G. Foster and Leon Kreitzman, *Circadian Rhythms: A Very Short Introduction* (New York: Oxford University Press, 2017).

154 Monarchs, too, respond to circannual rhythms: S. M. Reppert, "The Ancestral Circadian Clock of Monarch Butterflies: Role in Time-Compensated Sun Compass Orientation," *Cold Spring Harbor Symposia on Quantitative Biology* 72 (2007): 113–18, http://symposium.cshlp.org/content/72/113.full.pdf.

155 "They tolerate each other": Patrick Guerra, too, is a tolerant life-form. He spent many hours helping me explain this complex research in a way that is both (hopefully) accessible to the average reader and also technically correct.

155 monarch researcher Patrick Guerra: Guerra, now on his own, was a graduate student working in the lab of neuroscientist Steven M. Reppert. A substantial number of scientific papers like this one—"Neurobiology of Monarch Butterfly Migration," http://reppertlab.org/media/files/publications/are2015.pdf—are available via the Reppert

Lab website. Under the news/outreach category of the site's home page are several lengthy video presentations that explain this research in considerable detail.

166 Hugh Dingle and colleague Micah Freedman wanted to know more: "Wing Morphology in Migratory North American Monarchs: Characterizing Sources of Variation and Understanding Changes through Time," *Animal Migration* 5, no. 1 (October 2018): 61–73, https://www.degruyter.com/view/j/ami.2018.5.issue-1/ami-2018-0003/ami-2018-0003.xml.

166 evolutionary ecologist Martha Weiss has suggested: https://journals.plos.org/plosone/article?id=10.1371/journal.pone.0001736.

168 "My mind drifted, hypnotized": Rick Ridgeway, *The Last Step: The American Ascent of K2* (Seattle, WA: Mountaineers Books, 2014), 161.

168 evolutionary biologist Gerard Talavera: Talavera has an exceptionally well-maintained website with numerous articles about his research and with a number of helpful videos. You can access it here: http://www.gerardtalavera.com/research.html.

170 "No butterfly even hesitated": Hugh Dingle, *Migration: The Biology of Life on the Move* (New York: Oxford University Press, 2014),14.

Thirteen. Paroxysms of Ecstasy

171 "The eyes of butterflies are remarkable": Adriana D. Briscoe, "Reconstructing the Ancestral Butterfly Eye: Focus on the Opsins," *Journal of Experimental Biology* 211, part 11 (June 2008): 1805–13, https://www.ncbi.nlm.nih.gov/pubmed/18490396.

171 "The trade in rare butterflies": Matthew Teague, "Inside the Murky World of Butterfly Catchers," *National Geographic*, August 2018, https://www.nationalgeographic.com/magazine/2018/08/butterfly-catchers-collectors-indonesia-market-blumei/.

171 "the world's most wanted butterfly smuggler": Field Notes Entry, "Smuggler of Endangered Butterflies Gets 21 Months in Federal Prison," U.S. Fish and Wildlife Service Field Notes, April 16, 2007, https://www.fws.gov/FieldNotes/regmap.cfm?arskey=21159&callingKey=region&callingValue=8.

172 Response to color has been embedded: There are a number of texts on the subject of eye evolution. I used Thomas W. Cronin et al., *Visual*

Ecology (Princeton, NJ: Princeton University Press, 2014), https://academic.oup.com/icb/article/55/2/343/750252. Michael F. Land's authoritative *Eyes to See: The Astonishing Variety of Vision in Nature* (New York: Oxford University Press, 2018) is slightly more accessible to the lay reader.

173 It's visceral. Inextricably linked to survival: For an absolutely excellent discussion of how the brain processes color, Nobel winner Eric R. Kandel's *Reductionism in Art and Brain Science: Bridging the Two Cultures* (New York: Columbia University Press, 2016) is a must-read. The short and easily understood book is filled with visual information to back up the author's understanding of the link between art itself and the hows and whys of our attraction to these works.

173 wired into our psyches: A number of recent books address the connection between beauty and survival. In my reading, I included Richard O. Prum's *The Evolution of Beauty: How Darwin's Forgotten Theory of Mate Choice Shapes the Animal World—and Us* (New York: Doubleday, 2017). I also found Michael Ryan's *A Taste for the Beautiful: The Evolution of Attraction* (Princeton: Princeton University Press, 2018) to be very helpful.

174 inferior pathway, superior pathway: Kandel, *Reductionism.*

176 The common cabbage white: Kentaro Arikawa, "The Eyes and Vision of Butterflies," *Journal of Physiology* 595, no. 16 (August 2017): 5457–64, https://www.ncbi.nlm.nih.gov/pmc/articles/PMC5556174/.

177 modifying their responses to colors: "Color Vision and Learning in the Monarch Butterfly, *Danaus plexippus* (Nymphalidae)," *Journal of Experimental Biology* 214 (2014): 509–20, http://jeb.biologists.org/content/214/3/509.

177 Next, the scientists trained their monarchs: The Briscoe Lab, headed by entomologist Andriana D. Briscoe, has devoted a great deal of time to understanding how butterflies—and in particular the monarch—use their complicated visual abilities. For more information, you can go to the Briscoe Lab's website: http://visiongene.bio.uci.edu/Adriana_Briscoe/Briscoe_Lab.html.

Fourteen. The Butterfly Highway

182   Monarch Joint Venture, a collaborative group: https://monarchjointventure.org.

184   findings were posted on Journey North: https://journeynorth.org.

187   I had come to The Wilds: https://thewilds.columbuszoo.org/home.

193   the Tribal Alliance for Pollinators: https://tapconnection.org.

194   founder and head of Monarch Watch: https://www.monarchwatch.org.

Photo Credits

1 Greg Auger
2 Greg Auger
3 Matthew Lehnert
4 Matthew Lehnert
5 Ryan Null and Nipam Patel
6 Ryan Null and Nipam Patel
7 Greg Auger
8 Greg Auger
9 Greg Auger
10 Greg Auger
11 Greg Auger
12 Copyright Carol Komassa, Photographer
13 Greg Auger
14 Greg Auger
15 Greg Auger
16 Greg Auger
17 Greg Auger
18 Greg Auger
19 Greg Auger
20 Greg Auger
21 Holli Hearn
22 Holli Hearn
23 Ryan Null and Nipam Patel
24 Greg Auger
25 Greg Auger
26 Greg Auger

27 Greg Auger

28 Artwork courtesy of the National Park Service, NPS/HPCC/Rob Wood

29 Courtesy Albany Pine Bush Commission

30 Albany Pine Bush Commission

31 Greg Auger

32 Wikicommons

33 Image number B11797897 8 American Museum of Natural History

34 Image number b11797897 2 American Museum of Natural History